SASQUATCH

SASQUATCH

LEGEND MEETS SCIENCE

JEFF MELDRUM

FORGE®

A TOM DOHERTY ASSOCIATES BOOK / NEW YORK

A Forge Book
Published by Tom Doherty Associates, LLC
175 Fifth Avenue
New York, NY 10010

www.tor.com

Forge® is a registered trademark of Tom Doherty Associates, LLC.

Library of Congress Cataloging-in-Publication Data

Meldrum, Jeff.
    Sasquatch : legend meets science / Jeff Meldrum.—1st ed.
        p.   cm.
    Includes bibliographical references and index.
    ISBN-13: 978-0-765-31216-7
    ISBN-10: 0-765-31216-6
    1. Sasquatch. I. Title.
QL89.2.S2M45   2006
001.944—dc22

                                                    2006004267

First Edition: September 2006

Printed in the United States of America

0 9 8 7 6 5 4 3 2 1

Acknowledgments

This project is indebted to Doug Hajicek and all at Whitewolf Entertainment for undertaking the production of a thoughtful documentary that presented the evidence to a discerning audience and to the Discovery Channel for raising the bar in the treatment and consideration of this subject matter; to Michael Hsu for sharing our mutual desire to see a companion volume grow out of that creation and inviting me to author it.

The scope of this book exceeds the experience and expertise of any one person. I am grateful for the objectivity and open-mindedness of the featured scientists and their willingness to follow the data wherever it might lead them along a path that few have deigned to tread: George Schaller, Daris Swindler, Doug Divine, Bob Francis, Bill Taft, Carl Anderson, Reuben Steindorf, Andrew Nelson, John Bindernagel, Lynn Rogers, Loyd Benson, Jimmy Chilcutt, Henner Fahrenbach, and Craig Newton. To this list I would like to add the name of Richard Greenwell, cryptozoologist extraordinaire, who once asked me what I, as a primatologist, thought of sasquatch. Special recognition is afforded the several members of the North American Ape Project, especially Ron Brown, LeRoy Fish, Derek Randles, John Pickering, Rick Noll, Owen Caddy, Brian Smith, John Mionczynski, and Julie Davis.

As one who might be considered a member of the second generation of researchers pursuing this question, I must acknowledge the pioneering efforts of the first watch, who have dedicated so much of themselves toward the resolution of this intriguing natural mystery, and have so generously shared their hard-won experiences with me personally, in particular Grover Krantz, John Green, Bob Titmus, Roger and Patricia Patterson, Bob Gimlin, Wes Sumerlin, Bill Laughery, and Vance Orchard. Many members of the contemporary "Bigfoot community," too numerous to name individually, have contributed immeasurably through their shared experiences, photographs, casts, and samples. Numerous other professional colleagues engaged in scintillating discussions about the subject and kindly permitted the use of a number of the illustrations that are included herein.

A number of individuals and institutions have funded aspects of my own research directed at this question, both in the laboratory and in the field, including Brant Steigers, Ken and Linda Steigers, Jack Mayfield, Richard and Thomas Stepp, and the Idaho State University Research Committee.

Exceptional thanks are extended to George Schaller for graciously providing the foreword to this book and for lending additional expressions of encouragement. Also, to Jane Goodall for affording the time to read and endorse this book and to hazard an open position of acknowledgment of the value and need for this inquiry. Other valued colleagues who have read and commented upon the manuscript include Russell Mittermeier, Peter Matthiessen, Walter Hartwig, Roderick Sprague, Daris Swindler, Esteban Sarmiento, and John Fleagle. Significant encouragement was received in the early stages of this project from Peter Nevraumont. For their ongoing guidance and assistance, particular recognition goes to members of the editorial team at Forge: Moshe Feder and Denis Wong in editorial, and Barbara Cohen in legal.

Special acknowledgment goes to my wife, Terri, for support and indulgence of my preoccupations and to our six sons, Colin, Sean, Devin, Trevor, Kayd, and Brendon, for keeping me young at heart in spite of the encroaching gray.

Dedicated to the memory of the pioneering research of
Grover S. Krantz (1931–2002)

(Courtesy of Rick Noll)

Contents

Foreword

Some years ago, a project on mountain gorillas brought me to the attention of those in quest of sasquatch, yeti, and other apelike beings who are said to roam forests and mountains in various countries, from Congo, Indonesia, Russia, and Nepal, to the United States and Canada. I was already aware of the distinctive footprints, supposedly of yeti, photographed by Eric Shipton in Nepal in 1951, and later Edward Cronin and Jeff McNeely showed me casts of similar prints in 1972. Bhutan's Forestry Department has presumed yeti casts in its Thimphu office. Earlier my organization, the Wildlife Conservation Society, had been given yeti hair from Bhutan for DNA analysis (it was human hair). A Chinese team in Lhasa searching for the *yeren* (wild man) showed me photographs of its sleeping beds (they were nests of Asiatic black bear). John Green gave me a cast of a sasquatch footprint, truly a big foot, very different from that of yeti, and I have read his books as well as those of John Napier, Dmitri Bayanov, John Bindernagel, and others. I mention all this because sasquatch and yeti have walked through the forests of my mind for over three decades.

Naturally I am intrigued. I realize that the evidence includes hoaxes, delusions, and mistaken identities, prejudiced conclusions, and cultural legends of dubious value as testimony. It has been claimed that yeti footprints in snow are nothing but the melted tracks of hermits, bears, snow leopards, or whatever roaming the heights. But negative evidence does not disprove the yeti's existence. Still others view the large, hairy beings as mythic monsters gliding through our consciousness, lonely wanderers along the mysterious divide between humans and animals. Naturally we humans are mainly interested in ourselves, in our evolutionary history and our relatives along the road from this ethereal origin.

I look at the evidence with a naturalist's eye. I am neither a believer nor can I

Naturalist George Schaller observing wildlife in Tibet in 1985 (Courtesy of George Schaller)

reject all the evidence and conclude that a novel ape-like being cannot exist. Large unrecognized creatures may still roam remote forests. The saola, a primitive relative of wild cattle weighing 100 kilograms, was only discovered in the mountains along the Laos-Vietnam border in the early 1990s. The many sasquatch tracks need to be explained. Casts of such tracks, together with the cast of a body imprint with large heels clearly outlined, found in 2000, represent the most compelling evidence to date. The authenticity of films, especially the Patterson-Gimlin footage of 1967, has never been disproved. But there is still no proof. No bones, no skin, no conclusive DNA analysis from hairs. The question of existence remains open. But if even *one* set of yeti or sasquatch tracks were genuine. . . .

So far searches for these humanlike beings have been based on short expeditions, casual outings, or dependent on lucky encounters. A good field study of a species is based on months and years of work, of living in the wilderness to examine spoor and monitor shadowed forest trails, hoping for contact. At least new kinds of evidence might be obtained by such a much-needed extended effort. I am puzzled, for example, why so few large fecal piles have been reported along sasquatch trails. One would expect many from a bulky vegetarian with only rare access to meat, as shown by gorillas and giant pandas.

Jeff Meldrum is a scientist, an expert in human locomotor adaptations. In *Sasquatch: Legend Meets Science,* he examines all evidence critically, not to force a conclusion, but to establish a baseline of facts upon which further research can depend. His science is not submerged by opinion and dogmatic assumption. With objectivity and insight he analyzes evidence from tracks, skin ridges on the soles of feet, film footage, and DNA, and he compares it to that on primates and various other species. He disentangles fact from anecdote, supposition, and wishful thinking, and concludes that the search for yeti and sasquatch is a valid scientific endeavor. By offering a critical scrutiny, *Sasquatch* does more for this field of investigation than all the past arguments and polemics of contesting experts.

Humans stand alone, isolated monuments from a distant past. Perhaps we seek an evolutionary bridge to connect ourselves to something closer to us than the known great apes, to reveal more about how we became human. The apes offer only a distant glimpse. Possibly some day the yeti, sasquatch, or some other creature yet unnamed, will help to disclose more about the splendor and wonder of our own creation.

—George B. Schaller

SASQUATCH

INTRODUCTION

Doug Hajicek, nature film producer, took a break from shooting with his cameraman, and wandered near the shoreline of Selwyn Lake, nearly 800 miles north of Winnipeg, on the border of Saskatchewan and Northwest Territories, Canada. In the Arctic to film giant lake trout, they had flown into this remote location in the early 1990s. Along the isolated beach they encountered a crisp 17-inch footprint. The print was exceptionally clear and detailed, and excluding its enormous proportions, clearly humanlike in form, with distinct toes and a broad rounded heel. Some 40 inches farther ahead was another similar footprint, followed by another, and so on trailing alternately into the distance. This resembled no bear track. Besides, a polar bear hind paw measures only between 10 and 14 inches long. The hind paw of an Alaskan brown bear may reach a full 16 inches in length, but their range is generally restricted to the Pacific coastline. Grizzlies do range farther to the east but their foot is only about 10 inches long. Could it be an extraordinarily large and out-of-the-way Alaskan Brown, or an oversized grizzly, fishing for giant salmon? Hajicek weighed that possibility but he was familiar with bear sign from extensive documentary filmmaking with Lynn Rogers, the "man who walks with bears," and he knew that a bear track consists of a distinctive alternating pattern of hind and *forepaw* prints. The narrow interdigital pad of the forepaw is much abbreviated compared to the hind paw, to which is added an extended distinctly tapering heel pad. Whatever animal had left these tracks was walking upright, on hind feet only, and had struck off from the lakeshore in an apparently determined course with an impressive stride. Judging from the freshness of the tracks, it may even have been the film crew's arrival on the lake by floatplane that sent it on its way.

Hajicek's curiosity was piqued, and together with his cameraman, they followed the advancing line of footprints. For over a mile they traced the creature's enormous strides, before deciding that they didn't actually *want* to catch up to whatever behemoth had left the immense tracks clearly and deeply impressed in the frosty tundra soil. The men remained mystified over what could have been responsible for these prodigious footprints. They returned to the lake thinking that they could readily follow the tracks from the air over the relatively treeless landscape and perhaps overtake the track-maker. But the pilot of the floatplane refused to talk about the tracks and rebuffed their suggestions to pursue them, and so they gave up on the idea.

Hajicek was unfamiliar with sasquatch and therefore had no real concept of a giant upright ape upon which to hang the enigma of the footprints. The obvious and unavoidable fact that *some* unusual animal had made this impressive trackway continued to dog him, and his thoughts frequently returned to the scene of the discovery. The suggestion that someone might have intentionally hoaxed them at that precise spot beside a 70-mile-long lake in the middle of the Canadian wilderness, accessible only by floatplane, seemed absolutely nonsensical. Hajicek thought about it a great deal and eventually, with the proliferation of the Internet, encountered the Web site of the Bigfoot Field Researchers Organization (BFRO), and discovered that he was not alone in his experience. In fact, a surprising number of other people had discovered large inexplicable footprints in the wilds of North America. He also learned that the BFRO was compiling an electronic database of reports of sightings and footprints submitted by witnesses from all walks of life. A network of field investigators responded to and documented these reports, where possible, by interviewing witnesses and collecting corroborative evidence, or determining alternative explanations. Hajicek began to educate himself about the accumulated information concerning the history and nature of this hypothetical and strangely elusive primate. He was surprised and irritated that the public at large, and particularly the media, ignored the extensive evidence for the existence of this otherwise legendary animal. Being a filmmaker by profession, he thought what better project than to produce an informed documentary that dealt objectively with the data and explored the question of sasquatch with an open mind. The folks at the Discovery Channel concurred and so the concept of *Sasquatch: Legend Meets Science* was conceived.

The format of the documentary was a noticeable departure from the established formula for "monster" media. Instead of trotting out a series of sensational eyewitness accounts with interviews and dramatic re-creations, then "balancing" them with retorts by armchair skeptics and willfully ignorant scientific experts, Hajicek opted to let the data stand on their own. He would present the accumulated evidence on its own merits and enlist the expertise of scientists willing to evaluate it objectively and to pursue their analysis wherever it might lead them without prejudice. Several of these recruited scientists were previously unconcerned with the matter of sasquatch, but nevertheless, in the spirit of exploration, were quite willing to ply their skills to evaluate the evidence laid before them. Others harbored a longstanding interest in the subject, but had rarely spoken openly of it for fear of ridicule and a concern for their reputed credibility. A few of the scientists, like Dr. Bindernagel, Dr. Fahrenbach, and me, had already crossed paths with the evidence and were actively engaged in ongoing research into the matter, in spite of its unpopularity within mainstream science, and even among our own respective institutional colleagues.

Producer Doug Hajicek (center) during the filming of *Sasquatch: Legend Meets Science* (Courtesy of Doug Hajicek / Whitewolf Entertainment, Inc.)

The Internet has provided a novel and readily accessible forum for the exchange of ideas and information. Like Hajicek, other witnesses frequently submit reports of their encounters with sasquatch to the many sites on the Internet concerned with the topic. The BFRO was one of an overwhelming number of Web sites that one is confronted with when searching the Internet for information relating to sasquatch or Bigfoot. It was the one that captured Hajicek's attention and would subsequently cooperate during the development of the documentary concept. Many of the pioneering scientists Hajicek would work with were associated at one time or another with the BFRO. For a time the BFRO took the lead among a new generation of amateur and professional investigators. There were a number of organizations of various stripe, but the BFRO boldly touted the distinction of being "the only scientific organization probing the Bigfoot/Sasquatch mystery." A rather grandiose assertion perhaps, but, in so far as efforts were made by its investigators to adhere to the principles and methods of scientific research during the collection, handling, and evaluation of objective evidence, that standard was applied with varying success. Like any community, the BFRO was not without its volatile personalities, egos, strong wills, deep-seated opinions, conflicting agendas, and other controversies. However, in spite of intermittent lapses, there had been a degree of cooperation, collegiality, and professionalism among its collective membership.

Matthew Moneymaker, the founder and driving force behind the organization, recruited and sometimes rode roughshod over a lineup of amateur curators and investigators with varied skills and backgrounds. Affiliated with their ranks have been a number of credentialed scientists—primatologists, anthropologists, archaeologists, wildlife biologists, geologists, and engineers—who have pursued a professional interest in the matter. Some of the most dedicated field researchers, however, have had little or no formal training in the sciences, but often possess a vast experience in the outdoors and keen powers of observation and discernment.

To their credit, the BFRO investigators have been routinely critical of reports and taken pains to winnow the kernel from the chaff, concerning both potential evidence and would-be debunkers. Indeed, it is the proponent who is frequently responsible for refuting misidentified or misinterpreted evidence or claims. It must be appreciated that many of the individuals involved in the investigation lay claim to firsthand experiences that have effectively laid the question to rest for them personally. They are motivated by a conviction that eventually well-documented evidence will bear out their own experiences or convictions and resolve for them a vexing and persistent mystery.

As many youngsters then and now, I discovered an enduring fascination with extinct dinosaurs and prehistoric ape-men. I knew the author Jack London, not for his *Call of the Wild,* but for his less well-known novel, *Before Adam,* which explored the main character's dreamlike racial memories that nightly hurled him back to the vicarious experiences of remote forebears who lived in trees at the dawn of humanity. Growing up in the Pacific Northwest I eventually was exposed to the legend of sasquatch. At the age of eleven, I encountered Roger Patterson from a third-row seat in the Spokane Coliseum, where he was showing his captivating documentary film about "America's Abominable Snowman." Its centerpiece was the famous sixty seconds of jumpy footage of what he and partner Bob Gimlin claimed to have witnessed along Bluff Creek in the Siskiyou Mountains of northern California. The larger-than-life image of a Bigfoot deliberately striding across the screen made a lasting impression on a young and adventurous mind and served to reinforce my fascination with the evolution of the primates and primitive humans. Was this creature some "missing link" or some relic from a spent diversity of man-apes? I hadn't yet been indoctrinated concerning what *could* and what *could not* exist. The possibility that a giant humanlike ape, or some remnant apelike human, perhaps a relic of the Pleistocene Ice Ages, could have survived to the present in the remote corners of western North America, or elsewhere in the world, seemed to offer the prospect for a fascinating adventure in exploration. Patterson's dramatic film seemed to draw back the curtain on the legend, revealing what could be one of the most intriguing questions facing zoologists and anthropologists today. Does a giant upright ape inhabit our wilderness today? What, if anything, might it disclose about human history?

The Patterson-Gimlin film did not bring a speedy resolution to the mystery of the sasquatch, as Patterson and others had optimistically anticipated. In fact, it made very little lasting impact on the scientific experts of the day, in the absence of a body or some bones. The years that followed yielded no conclusive physical evidence, no type specimen required by hard science, and sasquatch remained stuck in the company of assorted legendary "monsters" and sundry occult subjects.

Very few physical anthropologists ventured to openly pursue a critical look at the matter. One exception was Dr. Grover Krantz, then a young anthropology professor at Washington State University, and not one inclined to shy away from a controversial idea, whether anthropological or cryptozoological (the search for "hidden" animals). He studied Patterson's film and concluded that it was in all probability authentic. He examined the tracks and concluded an unknown animal had left them; but more importantly, he persisted in thrusting the broader matter of the evidence for sasquatch under the noses of the "Scientific Establishment" as he came to rather critically refer to it. His colleague at the nearby University of Idaho, Roderick Sprague, editor of the *Northwest Anthropological Research Notes (NARN),* noted the lack of anthropological literature on sasquatch, and in a 1970 editorial invited responsible articles on the subject at a time when it was downright dangerous to one's career to do so. Over the next decade, a series of submissions was published in *NARN,* and these contributions were eventually assembled by Sprague and Krantz as a collected volume under the title *The Scientist Looks at the Sasquatch.* Without the early attentions of a few intrepid anthropologists, the subject might well have been altogether ignored by science and been relegated wholly to the realm of folklore and fantasy.

Krantz was an accomplished anatomist, and his detailed analysis of the Patterson-Gimlin film, his published evaluations of the accumulated

Dr. Grover Krantz, for many years the most visible scientist seriously investigating the sasquatch question, stands before a portion of his extensive cast collection of hominid skulls and sasquatch footprints at his home laboratory in Sequim, Washington, shortly after his retirement from Washington State University. (Courtesy of Rick Noll)

footprints, and his discovery of dermatoglyphics (skin ridge detail) on the casts of the sasquatch footprints could not be so offhandedly dismissed. However, dealing with "evidence" of such a controversial nature is not without its challenges and pitfalls, and Krantz became the target of criticism springing not only from skeptical professional colleagues but also from the volatile elements of the community of amateur Bigfoot researchers. In the end, Dr. Krantz did not live to see the mystery conclusively resolved, but he and others held the door ajar, creating the opportunity for later investigators to take an objective look at the matter.

My own early interests in apes and prehumans led me eventually into a career in academia, specializing in primate evolutionary biology. I focused on the emergence of human locomotor adaptations, especially our apparently singular trait of walking on two feet—bipedalism. Rather than center my investigations directly on the earliest initiation of hominid bipedalism, I have turned to the more recent pattern of emergence of the distinctive modern form of human walking, characterized by a striding stiff-legged gait and endurance walking and running.

Eventually, my path crossed that of Dr. Krantz. In 1996, during a visit with family in Boise, Idaho, I traveled with my brother Michael to Pullman, Washington, and the campus of Washington State University to examine Krantz's assemblage of alleged sasquatch footprint casts for myself. The day was spent pulling specimens from drawers and spreading them out on padded laboratory bench tops. There were casts from the Patterson-Gimlin film site; the enigmatic "cripplefoot" from Bossburg, Washington; a cast made by Sheriff Bill Closner from Skamania County, Washington, the first county with a formal ordinance prohibiting the killing of a sasquatch; casts from the Blue Mountains in southeastern Washington bearing skin ridge detail or dermatoglyphics; and many more. We compared and contrasted characteristics and discussed alternate interpretations of details of anatomy and foot function. We noted the repeated appearance through time of recognizable individuals residing in a given geographical region. Finally, we noted the dubious aspects of the more questionable examples and the evident hoaxes, although these seemed to be decidedly in the minority, contrary to my initial expectations. The singular opportunity for a firsthand examination of Krantz's collected series of footprints was extremely enlightening. The study of photographs can provide only so much, and then the investigator must examine the casts and the footprints themselves to truly appreciate the finer details and signs of animation. In 2001, Dr. Krantz formally passed the baton, and most of his cast collection was transferred to my laboratory at Idaho State University, where it joined the sample of casts I had also assembled in the meantime, totaling nearly two hundred casts. The Smithsonian Institution accessioned additional selected specimens from Krantz's collection, along with selections from his personal papers, and even his very own skeleton, now an enduring anthropological specimen.

A significant number of the casts in Krantz's collection, which also figure prominently in his notable book, *Big Footprints* (in its current edition known as *Bigfoot Sasquatch Evidence*), were accumulated from the nearby Blue Mountains outside Walla Walla in southeastern Washington. This region took on greater significance when I was asked to review a book, *Bigfoot of the Blues* (in its current edition known as *The Walla Walla Bigfoot*), written by a regional journalist by the name of Vance

Paul Freeman displays a cast of a sasquatch footprint that he collected in the Blue Mountains outside of Walla Walla, Washington. (Courtesy of Grover Krantz)

Orchard. For over a decade, Orchard had chronicled developments in the district in his newspaper column. Upon leaving Krantz's lab, my brother Mike and I paid an unplanned visit to Walla Walla, to call upon Orchard and a few of the people who figured prominently in his narrative: Wes Sumerlin, Bill Laughery, and Paul Freeman. Sumerlin was one of the last of the old-time mountain men, frequently horse-packing in the Blues. Bill Laughery was a former game warden now living in the Tri-Cities area. Paul Freeman had briefly been a patrolman for the U.S. Forest Service assigned to the Mill Creek watershed. It was on one of those patrols that he claimed to have encountered a sasqautch and subsequently found footprints that exhibited dermatoglyphics. Freeman had become something of a controversial figure, particularly among the community of amateur Bigfoot investigators, but I was eager to critically examine the original specimens that I had only seen copies of in Krantz's lab. Our unannounced visit found Freeman at home and he cordially invited us in to visit and to examine his footprint casts. The originals, with a few exceptions, were even more impressive than I had anticipated. I pressed him for more details about his casts, for examples of multiple casts from a single trackway, for circumstances of the finds, all the while attempting to size up the person, his reliability, and motivations. Shortly he turned to me and said, "You're obviously serious about this. Would you like to see some fresh tracks? I just found the first tracks of the spring earlier this morning."

I was quite incredulous and silently chuckled to myself. What a coincidence, I thought, to have fresh tracks on hand to show me, just like that. Could he have

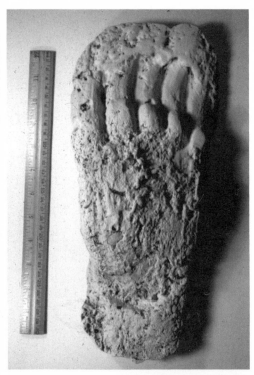

Examples of 14-inch footprints and casts documented by the author at Five Points, near Walla Walla, Washington, on February 18, 1996

learned of my last-minute decision to detour to Walla Walla and somehow hastily fabricated some tracks for my visit? It seemed unlikely, but either way, what did I have to lose? The three of us climbed into the truck and headed for the foothills. It seemed Freeman routinely drove the muddy mountain roads as early in the year as conditions permitted, looking for sign below the snowpack. That weekend in February, the melt-off had opened the lower foothill roads. On a restricted-access farm road he had found a long line of tracks in the wet ground. We pulled over, stopped the truck, and stepped out onto the muddy side road. A string of 14-inch tracks was plainly visible. Freeman's own tracks from earlier that day were also evident, but indicated he had merely walked alongside the tracks, pausing occasionally, presumably to inspect individual footprints more closely. Freeman repeatedly downplayed the tracks to me, saying they weren't that good and he wouldn't bother casting them, since he had seen much clearer tracks. However, what he considered shortcomings, to my eye were signs of their spontaneity and animation, although I still found the situation rather suspect due to the sheer coincidence, and I harbored lingering doubts about Freeman's credibility. "How could he have managed this?" I was silently asking myself as I surveyed the scene. Mike and Freeman wandered ahead as I began a closer examination, taking measurements and snapping photographs. The prints were 14 inches long by 5 inches wide. I knelt down close and could make out subtle patches of skin ridge detail, fading rapidly in the light drizzling rain. The tracks, whoever or whatever had made them, were fresh considering the weather conditions of the past several days, probably laid down during the preceding night or wee hours of that very Sunday morning. In some tracks the toes were extended and often the fourth and fifth digits hardly left a discernable imprint. In others the toes clearly curled over protruding stones; in still others the stones were pressed into the ground beneath the weight of the forefoot or heel, while still showing signs that a compliant foot had conformed to them. There were distinct tension cracks about the margins of many of the tracks—signs of dynamic compression rather than a forceful stamped impact. Several showed a speed bump–like ridge a little less than halfway along the length of the foot. This was clearly a pressure ridge marked by expansion cracks, which immediately brought to mind a picture I recalled of a track from the Patterson-Gimlin film site, and the corresponding cast I had just examined in Krantz's collection, which displayed a similar feature and dynamic details.

Then I came to a peculiar footprint that seemed to altogether lack a heel imprint. The step was on a slight incline and the foot had obviously slipped in the wet loamy mud. Distinct slide-ins were evident ahead of all five toes, which were sharply flexed and deeply impressed to gain purchase. The forefoot had pushed up a ridge of mud behind it, much more pronounced than in the other prints, but there was no heel imprint at all. It was similar to a person walking on the ball of his foot when going up an incline, except in this case the entire forefoot, not merely a ball, remained in contact with the ground. This indicated a greater degree of flexibility of the midfoot than is present in humans. The print was over two inches deep in the mud so that as the toes had splayed somewhat, the marginal toes had impressed into the sidewall of the track leaving a never-before-seen profile of the first and fifth digits. The three toe segments, corresponding to the three individual bones, the phalanges, of the little toe were discernable, while the big toe possessed only two segments. This is a subtle detail of skeletal anatomy that most people are quite unaware of. As the realization began to sink in that this could well be the track of a flesh-and-blood sasquatch, the hair stood up on the back of my neck.

Freeman rejoined me and described how it seemed that the tracks began about where we had parked the truck, made a hairpin loop in the soft soil of the adjacent plowed field, and ended once again along the side road precisely where the truck was parked. Again I thought to myself, how convenient for a hoaxer—simply don false feet in the back of a truck, jump onto the muddy road, trot out a truncated trackway, jump back into the back of the truck, doff the muddy false feet, and away you go. For the moment I kept these thoughts to myself.

The afternoon was waning and so we returned Freeman to his home and took our leave. I turned to my brother and said, "Even if these footprints are fakes, there is ample anatomy present. We can potentially learn a great deal from the incident, hoaxed or otherwise, by taking a closer look and making some casts." So after a quick trip to a hardware store for some supplies, we were back to the foothills. When we arrived, I said, "If we assume for a moment that these tracks are legitimate, then there has to be more sign up the road beyond the spot where the truck was parked and Freeman believed the tracks started and stopped." There was abundant surface along the road and adjacent fields to take tracks, but a thorough search beyond the truck turned up nothing. The only other option, besides a hoax, was that Freeman had read the sign incorrectly and the sasqautch had come and departed from the opposite direction. We began to flag each footprint methodically and soon recognized the point where the track-maker had made a hairpin turn, not out in the field, but near the truck, toward the road. We backtracked in the opposite direction along a brush-lined irrigation ditch and found more tracks beyond any sign of Freeman's footprints. Indeed, it appeared that whatever had made the tracks had come from the direction of the densely wooded Mill Creek that flowed out of the watershed, followed the cover of a brush-line ditch through the fields and nearby plum orchards, and was heading for the adjacent ridgeline that leads back to the watershed, when something, probably a passing car late on a Saturday night, prompted it to turn abruptly and retreat across the field toward the cover of the ditch, picking up its pace as it went. I was puzzled by the series of footprints immediately after the abrupt turn back away from the road. Every second right footprint was toed-out about 45 degrees from the line of travel. It wasn't until some months later, when I was examining my own tracks on an Oregon beach that I realized what had likely occurred. As I walked along the wet sand, I would occasionally glance back over my right shoulder to get a glimpse of the appearance of the tracks I was leaving along the beach. I quickly noticed that every time I looked back, my right foot toed out sharply in a manner very reminiscent of the tracks in Washington. Something caused the track-maker to turn abruptly and, with increasing pace, walk away from the road, glancing back every other step to assess the situation.

The plaster we bought was sufficient for seven casts and I tried to sample the variation evident in footprints depending on the conditions of the soil and the speed of walking or running. Some were shallower with the toes fully extended. Some were very deeply impressed, especially under the forefoot, in the softer soil of the fallow wheat field. I was especially interested in the "half-track" with the toe slippage and included a cast of it, and found another example of such. It was getting late and we both had commitments to fulfill the next day in Boise. I still harbored serious reservations about the whole set of circumstances and could not fully accept the situation on its face, but the more closely I considered the tracks the more intrigued I became. In hindsight, the incident had much more significance than I was prepared to acknowledge at the time. The drive home was punctuated with discussions of the possible meaning and implications of what we had witnessed, from Dr. Krantz's lab and cast collection, to Freeman's tracks in the foothills of the Blues. When we pulled into home, well past midnight, I was not inclined to retire just yet. Instead, I went to the garage sink and carefully unwrapped and washed the dirt from the seven casts we had retrieved. I lined them up and reexamined them, carefully noting the contours of the heel, the consistent protrusion of bony landmarks, the evidence of midfoot flexibility, the signs of articulation and obvious mobility evident in the toe impressions. I placed my little finger alongside the profile of the fifth digit in the peculiar half-track. It failed to cover it fully. The toes were relatively long, even for a 14-inch foot. The evident spontaneity and consistency of the tracks impressed me profoundly. Perhaps there *was* something to these footprints that deserved a much closer look. Who better to evaluate this evidence than someone long preoccupied with primate feet and the evolution of bipedalism? I began contemplating what it would involve to review and extend the line of study that Dr. Krantz had begun.

Some time later, when the opportunity was extended to participate in a sponsored expedition to attempt to collect new data in the field, I was keen to join. Richard Greenwell of the International Society of Cryptozoology had received funding from a foreign documentary film company to mount a four-week excursion into the Siskiyou wilderness of northern California. Our jumping-off point would be just a few tens of miles from the site at Bluff Creek where Patterson and Gimlin had their alleged encounter, captured on film in 1967. An additional five-man camera crew and four pack llamas accompanied the four-man research team. Our intended llama handler had to withdraw at the last minute and I was given a crash course in llama wrangling, which added a whole new and unanticipated dimension to the experience.

We had seismic sensors, night vision and call-broadcasting equipment, and early-model camera traps, at a time when few if any others were utilizing such technologies for this purpose. We discovered that such novel accoutrements presented a set of challenges all their own. On the first day in, we came upon an old set of tracks crossing a pass. They were heavily weathered and unsuitable for casting, but the discernable outlines of broad 13-inch footprints laid down in bipedal sequence were evident. After this tantalizing piece of evidence, little else was found over the next couple of weeks. The infrared triggers for the camera traps were set off by any interruption, animate or otherwise. When the morning fog rolled in off the coast it would set the cameras off in rapid succession, using up a roll

Scenes from the Six Rivers National Forest Expedition of 1997. Equipment was packed into the backcountry on llamas.

Author, Darwin Greenwell, and Richard Greenwell (Courtesy of Richard Greenwell)

of film in short order. We did get some intriguing responses to the calls we broadcast, but they were too brief and faint to record effectively. There was ample sign of bear scat in the areas we explored, but footprints of any kind were scarce, if not altogether absent, given conditions on the ground. Frankly, I was a bit surprised by how little of the ground surface lent itself to taking distinct tracks. The trails were rocky and a deep layer of dry duff often cushioned the forest floor. The creek beds were mostly rocky or densely overgrown with willows and alders. This was something I had not fully antic-ipated, and it goes largely unappreciated by most who seldom venture off-trail.

The camera crew departed on schedule after five days of shooting. We moved camp farther into the wilderness and hoped that with a fresh location, and fewer people about, we might have a better chance of encountering sign or detecting some-thing. As the fourth week began, our guide, Mark Slack, and I undertook an excursion about five miles farther along the trail, intending to then cut cross-country a farther cou-ple of miles to reach a cluster of off-trail lakes. We hoped that the shrunken lakes with exposed mud flats would provide more suitable substrate for tracking. The "trail" we followed was barely that since, as the Forest Service personnel had warned us, it had not been maintained for some years. We lost the indistinct trail a number of times as it crossed rocky outcrops, only to pick up portions of it again farther along. In many places it was strewn with numerous deadfalls, rendering it impassable had we brought llamas along. It was clearly a region rarely disturbed by human visitors. As we negoti-ated an east-west trending ridge the rocky trail was littered with a thin layer of duff still moist with dew. Abruptly, we came upon a series of fresh footprints 16 inches in length. Something had come up the slope, cut along the section of trail for a short dis-tance, and then continued upslope in the direction of a notch through the rocky ridge-line. The outline of the shallow imprints on the hard trail was unambiguous and the moist duff under the forefoot was scuffed back exposing the dry material beneath. The length of the step was over 42 inches on a moderate uphill incline. We scouted about, but what little sign we could distinguish off-trail was eventually lost in the rocky ridge.

About a mile farther on we left the trail, following a small tributary upstream, and made camp near a clear spring just below a ridge separating us from the lakes. The spring was surrounded by exotic-looking pitcher plants, giving the scene a surreal at-mosphere. Several bear trails were evident about the margin of the spring; their small quadrupedal tracks were in stark contrast to the footprints we had examined earlier that day. In the morning we climbed the ridge and descended through a vast boulder-strewn slope to the lakes. As we expected, the mud flats preserved many footprints, especially excellent bear tracks, some indicating cubs in tow, but disap-pointingly no tracks of sasquatch. We returned to the ridge just before sundown and

More scenes from the Six Rivers National Forest Expedition of 1997. The author's step fails to match that of a biped that left 16-inch tracks on the trail.

from that vantage point tried to imitate the recorded vocalizations we had broadcast from the base camp on preceding nights. Surprisingly, our voices seemed to carry quite a distance, echoing down the drainage in the still evening air. There was no response—no audible reaction that is.

As we reached camp a cold drizzling mist had engulfed the mountains and we huddled around the fire beneath our ponchos as we sipped miso soup left behind by the Japanese camera crew, and eventually retired to our respective bivy tents. I was awakened several hours later by Mark's urgent voice whispering "Jeff! Jeff, do you hear that?" As I roused from my slumber I could just make out a trailing cry some ways off in the night. I wasn't at all certain what I had heard, but I was certainly wide-awake now. I lay there as the minutes dragged on, straining to listen. Then came the sound of footsteps and popping brush circling our little camp, and a clacking sound, of rocks or perhaps teeth, in rapid succession. The clacking was promptly responded to by a second source of clacking on the opposite side of the camp. There was a pause and silence, and then a clackity-clack of Mark's pack frame bumping against the tree trunk it was leaning against. I struggled to extricate myself from my bag and tent, and heard Mark doing likewise. We found the pack, which had been wrapped in a poncho, standing uncovered with its flap unclasped and thrown back, its rifled contents hanging out in disarray. The dense mist obscured the moon and stars, and our flashlight beams did not significantly cut through the gloom. Whatever it was had simply slipped away into the fog. We lingered outside our tents for some time before returning to the shelter and warmth of our bedding. No sooner was I settled in than the footfalls returned. Their pace quickened and seemed to approach my tent with a rapid pad-pad-pad. The sound of footsteps passed along the side of my small tent and something bumped one of the poles, jostling it. "Mark?" I queried, wondering if he could be out there investigating something, but he responded from inside his tent in the opposite direction. I again scrambled from my confinement into the wet darkness and shot the beam of my flashlight along the side of my tent. Near the head of my tent was enough grass to momentarily hold a roughly 16-inch oblong impression, the grass just rebounding from the compressing weight of the tread. There were also punch holes in the boggy stretches below the spring where something had apparently stridden rapidly away from the campsite. The punch holes had slumped in with standing water and retained no detail.

The next day and night were uneventful except for incessant rain. We surveyed the immediate area for additional sign but found none, except for old washed-out bear trails. The next morning the rain let up and since our provisions were nearly expended, we broke camp and headed back to rejoin the others. Along the trail I hoped to get another look at what remained of the tracks we had encountered earlier along the ridgeline. They were in thick timber and relatively sheltered from the direct onslaught of the rains. We paused there for a respite and some refreshment. As I slung my pack off, a softball-sized rock sailed onto the trail a mere few feet away. A slight shiver crept up my spine. There was no high point nearby from which a rock might have been

Bivouac where a nocturnal visitor rifled through Slack's backpack and brushed against the author's tent.

dislodged by the rainstorm. Nor did it simply roll onto the trail from uphill. It had been airborne; it had been lobbed. For the first time on this excursion the hair on my neck stood on end; there was that subjective, but inescapable sense of being watched. In spite of the distinct feeling that something wanted us to move along, we determined to tarry and see what would happen. Nothing did and we eventually gathered ourselves up and continued on to the base camp, arriving late that afternoon and related our experiences to our companions.

That night I was again awoken by the sound of Mark's backpack clanging against a tree trunk. As I collected my senses, I called to Mark. Immediately there was a pad-pad of heavy footfalls running between our adjacent tents and the sweeping sound of something brushing along the length of my tent's rain fly. I emerged from my tent to see Mark's pack open again, this time the now-empty food bag was out and standing on the ground like an open grocery bag. The hard-packed ground had again taken no readily discernable tracks. In the morning we discovered that our food cache in the meadow had been raided. The twisted double garbage bags were standing open, but

Mark Slack, expedition outfitter (left), and author examine footprint sign.

there were no claw marks or tooth marks, and no scattered provisions that typically mark bear activity. Curiously, only a large resealable bag of flavored instant oatmeal packets was missing. The few remaining foodstuffs—instant potatoes, dried fruit, jerky, etc.—were left untouched. Such selectivity was also uncharacteristic of a scavenging bear, but had precedent in other reported sasquatch encounters.

These experiences in the Siskiyous were perhaps even more compelling than the footprints in the foothills of the Blues. We were in a remote region and had seen only two people besides our party in four weeks—some serious backpackers who had come through from the distant, more popular north end of the wilderness area. I was confident that no human was responsible for the events of those nights. I was equally convinced that we were not dealing with the antics of a marauding bear. The behavior was completely atypical of a bear in camp. There was something that had left 16-inch footprints, apparently walked on two legs, dexterously opened backpacks, rifled its contents without mark of tooth or claw, and accurately lobbed a rock in our direction. Just what that something was, remained hidden from sight throughout our excursion, but little doubt remained that I would have to pursue the question of sasquatch to its resolution, one way or another.

Hajicek's documentary project, *Sasquatch: Legend Meets Science,* set a new benchmark for serious media coverage of this intriguing cryptozoological mystery. It provided an opportunity to further my own research through discussion and collaboration with an assemblage of open-minded and inquisitive fellow scientists recruited to evaluate evidence in their respective fields of expertise. A colleague of Hajicek, Michael Hsu, suggested that a companion volume should be produced as well, in order to carry the dialogue to another level, and offered to take the lead in coordinating the effort. When Hsu approached me about writing such a book, I was very keen on the idea, having wondered about the possibility of a companion volume myself. While organized after the framework of the documentary, in this companion volume I delve even further into the background and history of scientific reaction to the persistent evidence than was possible within the time constraints of a one-hour television show. I share many of my own experiences, perspectives, and insights gained through nearly a decade of investigation and research. I also take the opportunity to set a baseline of what is known and to explore theoretical questions pointing to new directions of research. The latter may strike some as akin to the medieval debates about how many angels can dance on the head of a pin. But I suggest it begins to establish a framework for what is or is not *plausible* for the anatomy, behavior, and ecology of a reputed large primate. Perhaps it will bring us closer to a rational determination of what lies behind the legend of sasquatch.

THE SCIENCE OF HIDDEN ANIMALS: CRYPTOZOOLOGY

Whenever and wherever there have been travelers in exotic lands, or explorers of the murky frontiers of the known world, there have emerged accounts of exotic and sometimes fanciful creatures either actually encountered or merely heard tell of. The pages of medieval bestiaries depict curious and unfamiliar animals that seem to blur

the boundary between myth and reality. Many, once un-critically accepted, were eventually consigned by science to the realm of legend. Examples include the unicorn, the griffin, the manticore, and the mermaid. Others, it was determined, indeed had their basis in scientific reality, in the form of animals now considered rather commonplace in our age of televised natural history documentaries. These once fanciful, but now quite familiar animals include the leopard, the giraffe, the crocodile, and the mandrill.

Long before the present age of modern transportation and telecommunication, the eighteenth-century Swedish biologist Carl Linnaeus set out to catalog the whole of nature, laying the foundation for the modern scientific discipline of taxonomy. From excursions to the remote corners of the known world, Linnaeus's disciples returned not only with myriads of newly collected exotic specimens, but also with persistent reports of unusual and elusive creatures that stirred the imagination. Perhaps the most intriguing of these were creatures that

Fifteenth-century bestiary depicting fanciful animals, including a manlike creature, some of which are recognized today

Linnaeus's anthropomorpha, or manlike creatures, here rendered by Hoppius in the seventeenth century. Recognizable are the orangutan, the chimpanzee, and the baboon, but the identity, or even the reality, of the Troglodyte (at left) remains a mystery.

appeared to bridge the perceived gulf between human and animal. The possibility of the existence of such missing links was not only acceptable but anticipated, since naturalists of the day viewed the parade of life as a Great Chain of Being. The perfection of the Creator was expressed by the completeness of His creation. Therefore, the discovery of such intermediate species was inevitable. Among these rumored and rather bizarre-looking *Anthropomorpha,* or "man-shaped" animals, can now be recognized the baboon, the chimpanzee, and the orangutan. But a fourth manlike creature remains inexplicable. The legendary Troglodyte—a nocturnal, speechless, hirsute wildman—has never been recognized by science, although it occupies a key position in the traditional knowledge of ethnic cultures the world over.

It is noteworthy that the two most ancient examples of western literature, the epic of Beowulf and the epic of Gilgamesh, prominently involve the wildman figure in the characters of Grendel and Enkidu respectively. The image of the wildman carried into Medieval European iconography, adorning cathedrals, crypts, and heraldic emblems. In the East, ancient Mongolian literary sources also matter-of-factly depict such a wildman, the *zerlog khoon*. It is unceremoniously numbered among the commonly recognized wildlife of the Himalayas. To this day, questions remain regarding the identity of revered relics of the wildman enshrined in Nepalese monasteries. Is all this just a simple folk belief, or an animal species still hidden from science?

Western scientists briefly gave the matter of a relic wildman serious notice when mountaineers, exploring the roof of the world, came upon inexplicable tracks attributed to the abominable snowman, or yeti, of the Himalayas. The most famed were those photographed in 1951, by Eric Shipton and Michael Ward, in the Menlung Basin of Nepal. The large single footprint showing the outline of five rather oddly proportioned "toes" crisply impressed on a snow-covered glacier has been variously inter-

A Mongolian natural history of Tibet, written near the end of the nineteenth century, depicting a wildman

preted, and confusion still surrounds the associated photos that clearly depict a quadrupedal, rather than a bipedal track. Many, including the famous Sir Edmund Hillary, have chalked Shipton's tracks up to a runaway practical joke. However, as recently as 1997, Ward published a scientific paper revisiting the peculiarities and implications of the footprints from Menlung Basin, an unlikely move for the purveyor of a practical joke. Others have suggested that sublimation (the direct evaporation of snow at high altitudes) may have distorted the tracks of a more common animal, resulting in a "yeti" track. The case of the singular footprint would have benefited from supplementary photos of additional footprints in the trackway for the sake of establishing the consistency of its peculiar appearance and bipedal gait. In spite of the obscuring cloud of controversy, the highly publicized photographs of Shipton's footprint touched off the modern world's interest in the Himalayan wildman, spawning a number of sensational yeti expeditions during the late 1950s and early 1960s. The expeditions had various sponsorships: the London *Daily Mail,* World Book Encyclopedia, and the American oil tycoons Tom Slick and Kirk Johnson. Conclusive evidence was never obtained. However, as expressed by historian Daniel Boorstin, "To succeed in negative discovery—to prove that some mythical entity did *not* exist—was far more exacting and exhausting than to succeed in finding a known objective." This was especially true in the challenging circumstances encountered in the rugged remoteness of the Himalayas.

A series of letters appeared in the pages of *Science* magazine during 1957 and 1958, initiated by a brief article by Dr. William L. Straus, Jr., a physical anthropologist at the Johns Hopkins University. Straus was downplaying the probability of an unknown ape existing in the Himalayas, noting that the identification rested solely on footprints and verbal evidence. The footprints were summarily dismissed as those of misidentified bear spoor, which exhibit a superficial resemblance to human footprints. These had presumably been enlarged by sun and wind. The list of possible suspects venturing into the snow fields at high elevation to have their tracks enlarged by the elements grew to include yaks, lynx, snow leopard, wolf, ibex, deer, and langur monkey to name a few. Straus's explanation for the enigmatic footprints represented the generally adopted position. To this was added a comment by Dr. Robert K. Enders, zoologist at Swarthmore College in Pennsylvania, relating a firsthand experience with a track left by a man wearing snow sandals in Kasmir. The sandals were made from rough-woven leaves to protect the feet from the sharp ice. They wear out initially under the toes and then are eventually discarded. The enlarged and sometimes toed footprints left by such sandal wearers were suggested as a possible source for "yeti" tracks.

However, an exceptional response was forthcoming from Dr. Lawrence Swan, a native of India and world-renowned expert in high-altitude ecology. He took exception to the superficial and dismissive attitude of the foregoing scientists and pointed out, "The interpretation that tracks in the snow ascribed to the Yeti may be made by man is valid in some instances, but it is clear that footprints cannot logically be attributed to even the most solitary hermit when they are made in remote glaciated terrain at great altitudes where local inhabitants simply would not travel." He went on to say, "Perhaps the greatest difficulty with the bear theory, and the point most often disregarded in statements concerning Yeti tracks, is the fact that the high-altitude red bear of the Himalayas (*Ursus arctos isabellinus*) is found only in the western Himalayas, whereas the origin of the Yeti legend and the source of all 'genuine' Yeti tracks is in the eastern Himalayas. [Author's note: There actually are brown bears in the eastern Himalayas but not on the southern slopes.] There is a fairly striking faunal difference between these two regions, and it is not legitimate, nor is it good zoogeography, to attempt to discredit the legend on evidence obtained from the western Himalayas or the plateau of Tibet. The Abominable Snowman, presumably, has no business in these parts." As for the interpretation of volatile footprints he observed, "That the unique footprints may be the result of the high-altitude effects of evaporation and sublimation is not borne out by fresh Yeti tracks, where some detail of the foot is clear. High-altitude footprints do enlarge and may alter in shape, but this obvious alteration, which may surprise the casual traveler from the lowlands, is promptly recognized by an individual with experience in snow at high altitudes. It is not correct to assume that only the naive have seen the tracks, and it is equally erroneous to assume that the Yeti is only the imagined maker of all sorts of ablated footprints." Finally, while acknowledging the lack of conclusive evidence, he cautioned against summary dismissal of the evidence without due consideration, saying, "Whereas it is perhaps presumptuous to assume, at this time, that the Yeti is in reality some large anthropoid ape, it seems that this possibility has not been eliminated or sufficiently considered in the current arguments of the Yeti critics."

At this Straus retrenched, adopting a commonly repeated conservative posture, stating, "I certainly have never denied the possibility of the existence of an 'abominable snowman,' whether it be a giant ape or some other unknown creature. I am only adhering to a basic tenet of scientific procedure when I ask for something in the way of positive proof of its reality." It would seem that in some quarters, there is a distinction made between the conduct of science and the spirit of exploration. Some would wait at their lab benches and exert no effort to discover what, if anything, lies behind the legend. Yet, they are quite eager to engage the matter if someone else produces the conclusive evidence. Straus concludes, "If someone supplies me with the cadaver of an undoubted 'snowman,' I will be only too glad to dissect it and report, to the best of my ability, on the creature's zoological affinities."

Swan's optimism about the possibility that the yeti might exist waned considerably after he accompanied Marlin Perkins and Edmund Hillary to the Himalayas in 1960. On an opportune solo excursion he came upon a line of what he initially took to be "yeti" tracks in the snow. Upon closer examination he discovered, "The first prints were fairly good, although rather small by classic standards, but those further on seemed to change. Each footprint in the series was not sufficiently similar to its neighbor. I recognized that if I photographed only one choice track, I could astound anybody." As he moved along the line, the tracks became more distorted, until at last he stood over the clear imprints of a fox or small wolf. Swan concluded that sublimation could indeed sculpt from common tracks the enigmatic shapes of the mysterious yeti footprints.

Interest in the yeti was rekindled when, in 1972, biologists Edward Cronin and Jeffrey McNeely and members of their survey team discovered fresh nine-inch apelike tracks outside their tents high in the Arun Valley of Nepal. They had made camp on a ridge below a pass leading to a neighboring valley. During the night, something had made a detour from the ridgeline and investigated the camp, meandering among the expedition tents. The crisp bipedal footprints were evident in the snow, untouched by the morning sun and the effects of sublimation. The footprint spoor resembled a large ape's foot with an apparent divergent big toe and relatively long lateral toes. Backtracking revealed that the biped had ascended a steep slope through deep snow, without any apparent aid of its forelimbs. After the digression through the camp, the track

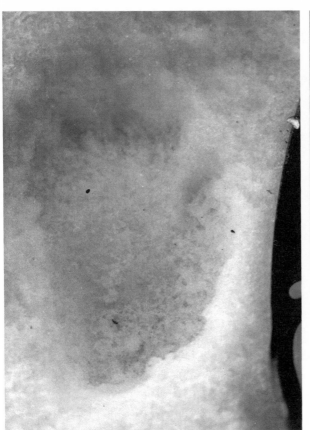
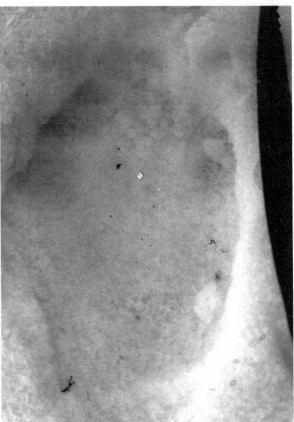
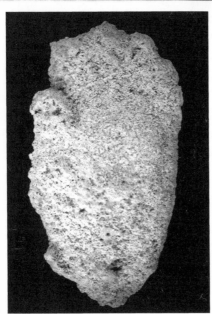

Fresh yeti footprints discovered and cast above the Arun Valley of Nepal by Cronin and McNeely in 1972 (Courtesy of Jeffrey McNeely)

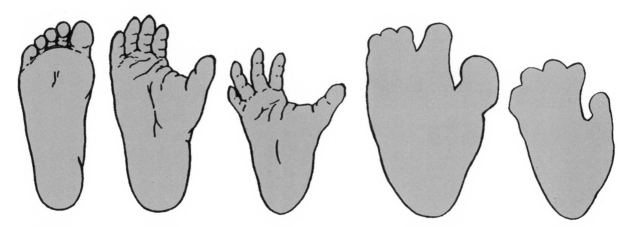

Comparison of the feet of (left to right) a human, gorilla, and chimpanzee, with the footprint outline of the Shipton and the McNeely footprints (drawn to approximate scale)

led back to the ridgeline and preceded over the pass to be lost among the boulder-strewn rhododendron thickets.

Upon examining the casts made of the footprints, Dr. George Schaller, who pioneered naturalistic studies of the mountain gorilla, noted that they "demonstrate a close resemblance to those of the mountain gorilla." Cronin concluded that the prints could not be attributed to any known animal of the eastern Himalayas, and therefore lent credibility to the theory that the yeti represents an unknown hominoid (ape) species. Even the report of this professional team of naturalists had little impact on a generally skeptical scientific community.

In 1993, four footprints attributed to the yeti were cast and today are on display in the Thimphu office of Bhutan's Forestry Department. They were collected in Hjage-hungla, Merak-Sakten, which is in east-central Bhutan. They are approximately 8 inches long, bear only four toes, and attest to the diversity of spoor that have been attributed to the yeti. To this day the maker of the unusual tracks in the Himalayas remains a mystery and continues to be the object of serious expeditions and amateur trekkers.

As early as 1825, the French naturalist Cuvier made the brash assertion that it was doubtful that any new large four-legged animals remained to be discovered. But shortly thereafter, he himself described a new species of carnivore found in the Himalayas—the red panda, *Ailurus fulgens,* Cuvier 1826. A long series of discoveries of large animals was to follow as naturalists made good on travelers' tales, and more especially, as they acted upon the natives' familiarity with local faunas. This approach was in large measure how new zoological discoveries were made. It was an established formula for investigating and discovering novel and exotic species. However, in

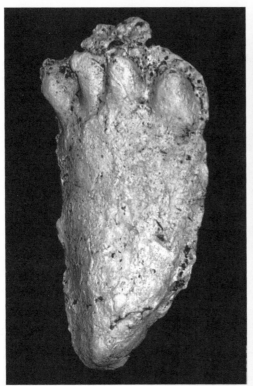

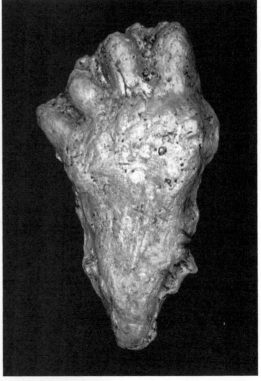

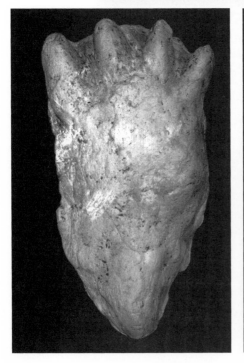

Casts of "yeti" footprints approximately 8 inches in length on display in the Thimphu office of Bhutan's Forestry Department. They were collected in 1993 at Hjagehungla, Merak-Sakten, which is in east-central Bhutan. (Courtesy of George Schaller)

more recent times this technique has fallen by the way and has been replaced by formal surveys that rely less on indigenous knowledge. Few scientists specifically search for rumored animals. Instead, most field biologists conduct broad taxonomic surveys to see what may turn up in their widely cast nets or narrow transects. An exception to this trend is Dr. Marc van Roosmalen, a Dutch primatologist who has a predilection for discovering new species of primates in the Amazon forest. By walking into a village with his eyes and ears open he routinely learns of unusual primates in the surrounding environs. His appreciation of native knowledge of local faunas often results in the description of a new species or even a new genus of Neotropical primate.

Contrary to Cuvier's premature pessimism, the pace of discovery has shown no signs of abating. When Linnaeus published *Systema Natura* (1735), his initial catalog of nature, there were nine thousand named species. Today, estimates of over 2 million species are described, although no all-encompassing tally has ever been undertaken. Serious estimates of how many species actually inhabit this planet range from 3 million to over 30 million! These newly discovered species are not limited to miniscule microbes or innocuous insects. As recently as 1929, a great ape was added to the list of known species, when the bonobo, or pygmy chimp, was found in the jungles of Zaire, Africa. During the past century, over two hundred additional species of primate have been discovered. In the Neotropics alone, twenty-four new species have been described since 1990, and at least ten more await formal description. Most recently, the prospect of a new ape, perhaps something intermediate to a chimp and a gorilla, has sent primatologists converging on the Congo in search of the so-called Bili (or Bondo) ape—with little more evidence to go on than some oversized footprints, nests, a few

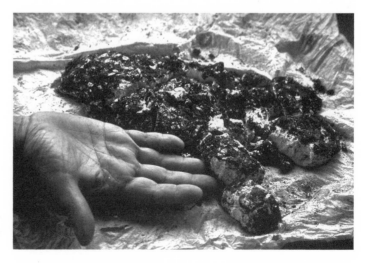

Cast (right) and outline tracing (left) of a footprint measuring over 12 inches long, attributed to the Bili Ape (Courtesy of Esteban Sarmiento)

strands of hair, and persistent native accounts of a large ape, which they call the "lion killer" due to its enormous size. Primatologist Shelly Williams, of the Jane Goodall Institute, experienced a close encounter with what she took to be four of these peculiar apes. They charged through the brush from less than ten meters away before they apparently realized she was not the quarry they anticipated. She described them as being huge, with a very flat faces, wide muzzles, straight overhanging brows, and grayish hair all over.

Indeed, persistent rumors of unrecognized apes, or less conventional wildmen, in addition to the yeti of the Himalayas, continue to emerge from forested mountain retreats the world over: from Mongolia, the *almasti*; from Indonesia, the *orang pendek*; from China, the *yeren*; from the U.S. and Canada, the *sasquatch,* or *Bigfoot,* to name but a few. These are not all described in similar terms and it has been suggested that several species may remain unaccounted for. For example, the yeti is reportedly smaller and more apelike with a footprint that exhibits a divergent great toe, while the sasquatch is much larger, more humanlike in posture, and has a footprint with the great toe aligned with the remainder. In spite of this multiplication of enigmas, or perhaps *on account* of it, many scientists are prone to retreat to Cuvier's smug presumption that the world is simply too small to still harbor any large undiscovered animals. Harvard paleontologist George Gaylord Simpson's reaction in 1984 is representative of this pervasive attitude: "As for the footprints and claimed sightings of the sasquatch, these have occurred in well-populated regions in British Columbia, Alberta, Washington State, Oregon, and northern California. It is simply incredible that so many educated people, including professional zoologists and anthropologists, should have failed to produce any objective evidence that yetis or sasquatch do exist." It seems that the majority of scientists are content to remain aloof, trivialize the probability of new discovery, or presume to discredit the witnesses and the evidence, leaving to others the search for the proof, the definitive type specimen. They passively challenge, "Show me the body."

As might be expected, into this relative void of passivity has stepped the tabloid media, eager to exploit the public's fascination with "monsters" and all things paranormal. The cult of the mysterious is a growth industry that mesmerizes the public, and is fraught with shades of pseudoscience and outright flim-flam. It seems that for many the tedium of everyday life in a technology-laden empirical world cannot be endured without an occasional whiff of the supernatural. Some will surely point to this book as yet another example of such exploitation. On the other hand, others have capitalized on the expansion of psuedoscience by commercializing an institutional skepticism. Behind a façade of critical thinking and objectivity, they set out to debunk anything and everything deemed unorthodox. Their object often appears to have more to do with selling books and magazine subscriptions than enlightening their readers about the diverse subjects they profess expertise in.

Unfortunately, the legitimate search for elusive animals has become embroiled in this mix of the mystical and pseudoscientific. Accounts of lake monsters and wildmen continue to be the stuff of sensational supermarket tabloids. Walk into a bookstore in search of reading material on Bigfoot and you will most often be directed to the occult section, somewhere between Bermuda triangle and crop circles. Few if any books on Bigfoot reside on the natural history shelves. The recent proliferation of television specials typically emphasizes the superficial aspects of the mystery rather than the substance behind the legend.

This peculiar situation amplifies the reluctance of conservative scientists to objectively consider the search for hidden animals, the field of cryptozoology. The challenge has been and continues to be to shift the reasonable questions of the existence and nature of these potentially unrecognized animals from the realm of the tabloid, squarely into the arena of zoological inquiry. Indeed, cryptozoology has gained increasing legitimacy in recent years. In January 1982, the founding meeting of the International Society of Cryptozoology was held in the Department of Vertebrate Zoology of the U.S. National Museum of Natural History, Smithsonian Institution, in Washington, D.C. Directed by a board of recognized zoologists from various subdisciplines, the society began publishing a newsletter and a refereed journal, and hosted meetings providing a venue for the reporting of serious research into the existence of rumored animals. Richard Greenwell carried the operation of the society, most recently housed in the International Wildlife Museum in Tucson, Arizona. However, since Greenwell's death in 2005, the future of the society is uncertain.

The unconventional discipline of cryptozoology received further acknowledgment when at the close of the twentieth century, the editor in chief of *Scientific American* prominently noted under the title, "Unexpected Thrills," that "zoology has been rocked during this decade by the capture of several large mammal species, some new to science, others that had been thought extinct, including the Tibetan Riwoche horse and the Vietnamese Vu Quang ox. The pace of these discoveries is astonishing . . ."

In a somewhat different vein, i.e., paleontological, but with clear implications, is the steady accumulation of fossil finds providing a revised notion of the pattern of hominoid evolution and the recent survival of some species of ape and early human. For example, the revised dating of *Homo erectus* fossils from Ngandong, Java, suggest these hominins survived in Indonesia to sometime between 53,000 to 27,000 years ago, much later than previously thought, and were likely contemporaneous with modern *Homo sapiens*. It should come as little surprise then, to those who have acknowledged the apparent bushy pattern of hominin evolution that significant new discoveries remain to be made. However, even that intellectual concession can hardly forestall the excitement and wonder over the discovery of *Homo floresiensis*, a new species of hominin in the Pleistocene of Asia, which apparently stood a mere 1 m in height and has a cranial capacity of only 380 cm^3. Even more intriguing is the geologic age of the find. The oldest remains date to 94,000 years ago, but some of the fossil material could be as young as 13,000 years old.

The holotype is a partial skeleton dubbed LB1 for the site of Liang Bua, a limestone cave on the island of Flores, in Indonesia. The partial skeleton exhibits a mosaic of primitive traits. The small skull is unquestionably that of a biped, with noticeably reduced facial height and projection. The orbits are large and rimmed above by arched brows. The forehead is sloping and the chin is receding. The lower extremity is well represented by a pelvis, thigh and leg bone and tibia. These are notably primitive and resemble those of australopithecines and early *Homo*. The shin bone or tibia is especially comparable to a chimpanzee in some features and distinct from modern humans. The crural index (ratio of leg to thigh length) is unexceptional at 84 percent. Nothing can be said about the relative length of the lower extremity since the vertebral column is insufficiently represented and the stature estimates were based on femur lengths for pygmy humans. The only insight into the upper limb comes from an unassociated and older radius. Although suggested to be proportional to the dimensions of the type specimen, the radius is substantially longer than human proportions would dictate. This prompted one commenter to speculate that *Homo floresiensis* was capable of an arm-swinging type of locomotion in the trees. After all, when sharing an island with giant carnivorous lizards, the ability to retreat to the tree tops might serve one well.

There are assumptions inherent to the working hypothesis that the inferred stature of LB1 is the result of insular dwarfism as proposed. To date there is no fossil evidence of full-sized *Homo erectus* or any other putative ancestor on the island. In fact there is meager morphological evidence that compellingly links LB1 to *Homo erectus* other than shared primitive traits and temporal proximity. Indeed very little formal discussion has been offered regarding the alternate hypothesis, which was acknowledged by researchers in press interviews, that *Homo floresiensis* arrived on the island already small-bodied. Perhaps this is a small relic hominin with reduced prognathism, a trend towards megadontia, reduction or loss of third molars, molarization of the lower fourth premolar, exhibiting extensive dental wear suggestive of a coarse diet. While LB1 has been identified as a female on the tentative basis of pelvic morphology, no mention or discussion is made of the potential for sexual dimorphism in canines or body size in this new species. The former is potentially hinted at by the indication of a diastema between the upper incisors and canines in LB1; the latter might account for the disproportionate length of the unassociated radius, in which case speculations about arm swinging are premature.

The stature is estimated based on height correlations to femur length in human pygmies. This method assumes *a priori* humanlike limb proportions. It should also be considered that the hindlimb may be relatively short in relation to torso length, as is the case in australopithecines or pongids. The body mass estimate from the assumed humanlike stature (28.7 kg) is considerably less than the estimate based more directly on femur cross-sectional area (36 kg), a difference of about 25 percent. This suggests the possibility of a larger stature for LB1 or a quite robust nonhuman body build, or a combination of both. The larger mass estimate places the relative brain size well below that for *Homo* and within the range for australopithecines and chimpanzees. This raises questions about whether *Homo floresiensis* was responsible for the manufacture of the tools found at the site. Described as "miniature" artifacts by some press, subsequent reports cite expert suspicions that the tools are too large to be associated with the diminutive *Homo floresiensis*. It was reported that incomplete hand and foot skeletons were recovered, however these were not described. Perhaps the anatomy of these appendages will shed further light not only on tool-making capabilities, but also on the manner of its bipedalism.

It appears that assertions that *Homo floresiensis* falls firmly within the genus *Homo* are presently overstated and run counter to some current trends in the application of that nomen. The association of stone tools appears open to some question and should not serve as the linchpin in a case for bestowing human status on these otherwise primitive hominins. The numerous hyperboles about rewriting textbooks of human evolution or rethinking what it is to be human are vacuous, but that distraction should not diminish the significance of this find.

In an editorial column in *Nature,* Henry Gee, naturalist and senior editor, boldly explores the most *obvious* potential implication of this discovery—that the tales related by regional natives of little hairy people may indeed have a basis in fact. For centuries both natives and Europeans have reported encounters with a short, but powerful ape with a rather humanlike face that walks on its legs like a man. In Indonesia it is called the *ebu gogo*; in Sumatra and Malaya it is the *sedapa* or the *orang pendek*. The latter translates as "short man." Gee suggests that in light of this discovery, efforts to search the jungles of Sumatra for the elusive *orang pendek* "can be viewed in a more serious light." That the evidence demonstrates that *Homo floresiensis* survived until very recently "makes it more likely that stories of other mythical humanlike creatures such as yetis are founded on grains of truth."

In the 1920s, interest in the ape-man of Sumatra was piqued by the report of a Mr. Van Herwaarden, who described an encounter with an orang pendek while hunting on the island of Palau Rimau. The creature stood 1.5 m in height, walked erect on two legs, with arms reaching to just above the knees, and was covered with dark hair excepting its face. In spite of its apish qualities, Van Herwaarden was afraid to shoot it because it otherwise looked so human. However, an ensuing series of misidentified bear tracks and hoaxed monkey corpses quashed any further serious consideration of the matter. In 1969, John McKinnon, a British naturalist observing Bornean orangutans, came across distinct humanlike footprints along a muddy trail. The prints were six inches long by four inches across and rather triangular in appearance. The toes looked quite human, as did the tapering heel, despite the remarkable shortness and breadth. McKinnon asked his boatman what made them, who promptly replied "*batutut.*" Named for its drawn out plaintiff call, the batutut was described as a type of ghost, a shy nocturnal creature about four feet tall, which walks upright like a man and has a long black mane. Later, in Malaya, McKinnon saw some casts of footprints even bigger than those he had seen in Borneo, but he recognized them as definitely having been made by the same animal, which there was called the orang pendek. These he compared closely to drawings and photos of similar footprints from Sumatra. Despite some similarities, he concluded that the footprints were too large to have been made by a sun bear.

Debbie Martyr, a journalist and conservationist, has pursued the orang pendek since 1989, even catching a fleeting glimpse of the furtive ape. Descriptions by Malayan natives are very consistent: 1–1.2 m in height, very strong with broad shoulders and long arms, short legs, covered with dark gray hair (or black flecked with gray), prominent arches over wide-set eyes, small mouth with broad central incisors and prominent canines.

Historian Caty Husbands has spent several years researching the island of Flores. She recorded accounts of contemporary interactions between the natives and the ebu gogo. She points out that if *Homo floresiensis* died out 13,000 years ago as has been suggested, then the stories of the ebu gogo illustrate the power of oral history. But if her insistent native informants are correct, then the creatures lived alongside humans until much more recently.

In spite of the accumulation of evidence Henry Gee notes complacency by science when dealing with the prospects of relic species of primates or even hominins. The purported evidence for the orang pendek, including foot-prints and hair samples that have defied identification, are now given greater credence in the wake of the discovery of *Homo floresiensis,* asserts Gee. It may turn out that the diversity of hominins was always high and might not have been entirely extinguished. Perhaps there is still room for yetis and their ilk on our ever-shrinking planet. "Now, cryptozoology, the study of such fabulous creatures, can come in from the cold," Gee says.

Of course, it's one thing to entertain the possibility of discovering new species in far-flung corners of the globe, but another matter to suggest that a large unknown ape may be lurking in one's very own backyard. Is it actually so incredible to consider that an ape may exist in North America?

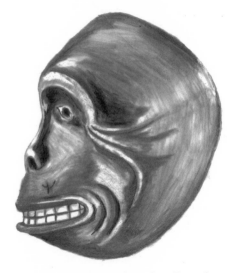

Tsimshian ceremonial mask collected in northern British Columbia about 1914 by G. T. Emmons

Although field zoologists have long recognized the wealth of knowledge held by the indigenous peoples regarding the local animals, little attention has been afforded the traditional beliefs of Native Americans about sasquatch, except by cultural anthropologists, who chronicle the folklore and mythology of native societies. The first inhabitants of North America were aware of these creatures long before Columbus arrived. They depicted them in their art and ceremonies along with animals more commonly known to the western world. Ceremonial masks and totem figures depicting the wildman display surprisingly apelike features, considering there are no apes in North America to serve as models for these effigies—at least none recognized. Some of these are quite stylized, while still retaining common identifying features. One artifact that is exceptionally realistic is the Tsimshian monkey mask from northern British Columbia collected in the early 1900s. Housed in the Peabody Museum at Harvard University, it is described in the accompanying documentation as representing "a mythical being found in the woods and called today a monkey." The artifact exhibits common anthropoid features, such as prominent brow, flat upturned nose, toothy grimace, and projecting chinless lower face. The only feature at odds with extant great apes is the lack of projecting canine teeth. Interestingly, the jaws of the presumed extinct *Gigantopithecus,* discussed in a later chapter, have relatively nonprojecting canines that wore even with the tooth row.

The motif depicted in the Tsimshian mask is repeated in prehistoric carved stone heads from the Columbia River Gorge. Owen C. Marsh first mentioned these in an address to the American Association for the Advancement of Science in 1877. He said, "Among many stone carvings which I saw there [Columbia River] were a number of heads, which so strongly resemble those of apes that the likeness at once suggests itself." Another observer commented in 1886, "Where the Indians of this region obtained the idea of such perfect baboons [Author's note: Historically, the term *baboons* referred generally to anthropoids, i.e., monkeys and apes.] is a mystery . . ." Renewed attention was directed to the heads by anthropologist Roderick Sprague, who discussed the heads and their potential significance at a conference on sasquatch held at the University of British Columbia.

The various tribes across North America have attached their own names to the entity. These names number more than sixty, but most generally make reference to a "wildman of the woods." In the 1920s, Canadian journalist J.W. Burns coined the term *sasquatch* as a common denominator for the myriad of native names. Sasquatch derives directly from the word "*sésquac.*" The original word, in the Stó:lõ dialect of the Halkomelem language, is used by the Coast Salish Indians of the Fraser Valley and parts of Vancouver Island, British Columbia. Other names describe particular behaviors associated with the animal, such as eating clams or shaking trees, thus reinforcing the impression that the natives are describing an animal they have actually encountered, and of whose habits they have firsthand knowledge.

Early encounters between sasquatch and Europeans in America date to the 1800s. An example is found in a letter to the *Antioch* (California) *Ledger,* in 1870. A hunter recounted that upon returning to his camp, he repeatedly found the ashes and charred sticks from the campfire scattered about. The ground in camp was hard and no tracks were evident, but a search of the vicinity revealed a set of large barefoot tracks nearby. In order to discover the identity of the visitor, he remained nearby the next day. From a hidden vantage he sat, overlooking the camp. In his own words he said, "Suddenly, I was surprised by a shrill whistle, such as boys produce with two fingers under their tongues, and turning quickly, I ejaculated, 'Good God!' as I saw the object of my solicitude standing beside my fire, erect and looking suspiciously around. It was the image of a man, but it could not have been human. I was never so benumbed with astonishment before. The creature whatever it was stood fully five feet high and disproportionately broad and square at the fore shoulders, with arms of great length. The legs were very short and the body long. The head was small compared with the rest of the creature, and it appeared to be set upon the shoulders without a neck. The whole was covered with dark brown and cinnamon-colored hair, quite long on some parts, that on the head standing in a shock and growing down close to the eyes . . . As I looked he threw back his head and whistled again."

Cast of a 16-inch footprint and the footprint itself, photographed and cast by reporters from the *Humboldt Times,* at Bluff Creek, California, in 1958

Such reports were familiar to those frequenting the mountainous forest regions, but were not common knowledge elsewhere. In the late 1950s, however, public attention in the United States was galvanized when an unusual story captured the journalistic attention of the newspaper wire services. New roads were encroaching upon the remote forests of northern California. Apparently, bulldozers represented an object of curiosity to one particular resident creature, as abandoned campfires had been in the previous century. Occasional nighttime forays to investigate the previous day's earth-moving activities were attested to by lines of tracks along the roadside and encircling the idle equipment. Construction workers dubbed the nocturnal visitor "Bigfoot," for the oversized humanlike footprints it left in the freshly turned earth.

When a bulldozer operator named Gerald (Jerry) Crew, a man with a reputation for honesty and levelheadedness, showed up in town with a plaster cast of one of the 16-inch footprints, Bigfoot made headlines at home and across the country. Generally, scientists took little notice and made no serious efforts to investigate the reports. An exception was Ivan Sanderson, a noted zoologist and animal collector from New York. He examined the casts and discussed the California Bigfoot in his encyclopedic treatment of wildmen around the globe, *Abominable Snowmen: Legend Come to Life.* At first he was incredulous—"It is very well to have abominable creatures pounding over snow-covered passes in Nepal and Tibet . . . but a wildman with a 16-inch foot and a 50-inch stride tromping around California is a little too much to ask even Californians to accept." However, based on the evidence, he concluded that an unrecognized animal *was* responsible for the footprints, although that judgement by him was not taken seriously in most quarters.

Even the seminal work of noted primatologist John Napier, another rare scientist who was willing to consider the evidence with an open mind, had little lasting impact on the academic community. After a review of a sample of the evidence, Napier drew a conclusion and revealed what contributes to the continued resistance to this proposition. He said, "One is forced to conclude that a manlike life-form of gigantic proportions is living at the present time in the wild areas of the northwestern United States and British Columbia. If I have given the impression that this conclusion is—to me— profoundly disturbing, then I have made my point. That such a creature should be alive and kicking in our midst, unrecognized and unclassifiable, is a profound blow to the credibility of modern anthropology." These are the expressions of a few intrepid scientists who were willing to consider the possibilities in light of the data, while still wrestling with the implications of their interpretations and conclusions.

Are these reports more than just stories? Can the persistent and remarkably consistent accounts by eyewitnesses from all walks of life—experienced hunters, police officers, foresters, park rangers, wildlife biologists, and field geologists, to name but a few—simply be dismissed out of hand as the product of mass hysteria or delusion? Do mere "stories" lay down tracks, shed hair, void scat, or leave sign of foraging? It is one thing to casually dismiss a report from the security and comfort of one's armchair. It is quite another thing to look into the face of an experienced outdoorsman and tell him he is mistaken or worse yet, a liar. It is yet another matter, a betrayal of scientific principles, to decline to examine and consider the evidence because after all, such creatures as the sasquatch "*cannot* exist, therefore they *do not* exist," so why be bothered with questionable "evidence." And yet, such is the atmosphere that has prevailed in scientific circles.

However, my sense is that, of late, conditions and a few attitudes have noticeably shifted. Open expressions of support and encouragement from some prominent researchers, not least among them the likes of George Schaller, cannot be idly dismissed by serious and informed scientists, let alone armchair skeptics. It seems that several factors may have contributed to this shift. First, a great deal has been learned about the fossil record of human evolution. There is a new outlook on the adaptive role of bipedalism, or the habit of walking on two feet, that no longer simply equates bipedalism with humanness. Second, the pioneering efforts over the past several decades of dedicated naturalists and field primatologists such as George Schaller, Jane Goodall, Dian Fossey, and Biruté Galdikas, and many more that have carried on their legacy, have yielded a clearer understanding of the natural behavior and diversity of the known great apes. Third, a new generation of anthropologists and zoologists has grown up with the contemporary notion of sasquatch (even explicit discussion of it in some widely used anthropology texts), rather than having had it dropped abruptly into their unsuspecting laps. Shifts in paradigms, regardless of the evidence at hand, frequently require the rolling over of a generation in order to take root in acceptance.

Whether it exists in reality or not, the sasquatch has certainly become a part of the North American landscape, both culturally and scientifically. Bigfoot and sasquatch are the symbol of monster trucks and hockey teams; they sell everything from pizza to athletic shoes. Witnesses are less afraid of ridicule should they decide to share their encounters with others. The question of its existence and place among primates has been discussed at professional conferences and exhibited in the halls of prominent museums. A growing number of scientists no longer perceive the sasquatch as such an *extra*ordinary possibility. For some, it boils down to a question of the *probability* or likelihood that such an animal could exist unconfirmed at this time in this place.

Here, in this book, the *plausibility* and *probability* of a North American ape has been put to the test of science as never before. Diverse evidence has been and continues to be scrutinized by experts from varied scientific disciplines: paleontologists, primatologists, anthropologists, forensic examiners, image analysts, biomechanists, naturalists, trackers, animators, kinesiologists, statisticians, and molecular biologists. These experts have applied their skills to objectively evaluate the available data and have drawn their own conclusions about the evidence. What has been revealed, and are the conclusions sufficient to give pause to the skeptics, or to the believers? Find out as the legend of sasquatch meets science.

2

WOODEN FEET AND FUR SUITS:
RAY WALLACE AT BLUFF CREEK

The phone rang, interrupting my grading of term papers. A voice introduced himself as Bob Young, a staff reporter for the *Seattle Times*. He was writing an obituary for Mr. Ray Wallace, who passed away on November 26, 2002, in Centralia, Washington. I rec-ognized the name. Wallace had been the contractor on the Bluff Creek road job in Humboldt County, Cal-ifornia, back in 1958. His brother Wilbur was the fore-man. After enormous humanlike footprints began turning up intermittently at the construction site, one of the seasoned Caterpillar operators, Jerry Crew, cut an outline of the footprint from a piece of cardboard and took it into town to the local taxidermist and tracker, Bob Titmus. Titmus could not identify what had made the track, but supplied Crew with instruc-tions and materials to create a cast of a footprint using plaster. When Crew returned with a clear cast over 16 inches long, he and it were featured on the front page of the *Humboldt Times* and the story was picked up by the wire services. "Bigfoot," as the mysterious track-maker was dubbed, became a household word—an all-American version of the recently popularized Abominable Snowman, or yeti of the Himalayas.

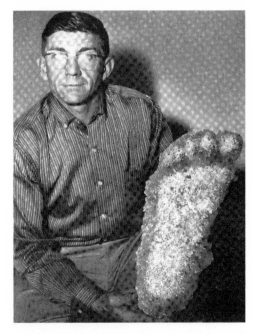

Catskinner Jerry Crew holding cast of a 16-inch footprint found around his Caterpillar at the Bluff Creek road construction site in October 1958 (Courtesy of *Humboldt Times*)

Ray Wallace was something of a colorful character and was known as a practical joker and "prankster." If the sensational footprints were hoaxes, it was generally assumed that Wallace was likely behind it. In fact, the local county sheriff openly accused him of it, even though Wallace was out of the region at another job site when the footprints first turned up. When he returned Wallace was reportedly incensed by the accusation and vigorously denied the charge. "Who knows anyone foolish enough to ruin his own business, man?" Wallace retorted. Reportedly fifteen men had quit their jobs after the discovery of the giant footprints. "I've got three tractors sitting up there without operators, man, and the brush-cutting crew has all quit. It just doesn't make sense."

Given the remote location, the recurring enormous tracks were a bit unnerving, even for the intrepid road crew. When Wilbur "Shorty" Wallace, Ray's brother, reported finding heavy equipment moved about, 50-gallon drums of diesel fuel, 700-lb spare tires, large steel culverts, tossed about without explanation other than the associated giant footprints, the crew became even more disconcerted. Was Shorty just having fun with the men? Ed Patrick didn't think that was the case. He described to me a curious incident. It seems the large steel culverts were delivered to the site in fours, bundled with heavy gauge wire cable about 3/16 of an inch thick. When the cable was cut loose it was coiled up and bundled, weighing over 100 lbs and left lying until it could be hauled away. Patrick and Titmus were scouting for sign of the track-maker up the mountain above the construction camp when they came upon one of these unwieldy bundles of cable, which had been carried a considerable distance up the mountainside off-road. Patrick, who was initially skeptical about the whole affair, couldn't imagine why anyone would have hauled that heavy bundle through rough steep terrain to such an out-of-the-way location, even if someone were *capable* of doing it.

If this were just a prank, why would the Wallace brothers intentionally or otherwise sabotage their own operation? One former logger, John Auman, suggested that Wallace planted the giant footprints around construction equipment to scare thieves away. "If your rig was parked overnight, you might as well figure it would have no tires in the morning," Auman said. "That's why this all started." When the prank attracted national attention, Wallace kept his role to himself, Auman speculated, concerned he'd get in trouble with the sheriff. Auman acknowledged, "He didn't say he done it, but I knew he did." Curious logic, since a thief would hardly notice the footprints at night. Ed Schillinger, a stake setter on the work site, said some men lived in camp at the site and there was always someone there, nights and weekends. It was more than an hour's drive over the mountain on a precarious dirt road to get to the site and there was only one way in and out. Thieves or hoaxers would have to go right through the camp to get to the construction site where the tracks were made. And if

Examples of Ray Wallace's hoaxed Bigfoot footprints and spherical stones (geodes) that he claimed were fashioned by Bigfoot to dispatch game (Courtesy of Rick Noll)

the ploy was directed at the thieves, why was the crew left unawares, resulting in more than a few being scared off the job? And wouldn't Wallace expect the cooperation of the sheriff in the prevention of acts of larceny that threatened the completion of his contract?

It has also been suggested that the hoaxing was actually about having a great excuse for failing to fulfill the contract. Loren Coleman says it might have been a case of "Bigfootgate." The subcontracted work had apparently fallen behind schedule and perhaps if Wallace could show that he could not keep a full crew on site then he could obtain an extension. In the end the Wallace brothers abandoned the contract and reportedly lost a sizable sum. In a letter, Ray claimed to have lost $40,000 on that road job as a result of the incidents.

Wallace himself offered a different rationale for his alleged footprint hoaxing. Loren Coleman relates that a letter from Wallace explained that when bear hunters with hounds began hunting for Bigfoot, Wallace became concerned for the big guy's welfare. He told the hunters he had made those tracks, but when he couldn't produce the fake feet, the hunters continued their search. He determined that he would have to get some wooden feet from a friend of his (Rant Mullens) who was particularly handy with a broad axe. He paid fifty dollars for a pair of carved feet. However, Michael Dennett, a contributor to the *Skeptical Inquirer,* reported that according to Mullens, the first set of carved feet he supplied to Wallace were made in 1969, nearly eleven years after the events at the Bluff Creek construction site. When he showed these to the hunters, they reported him to the Humboldt County sheriff. Wallace goes on in the letter to give the reason for his concern for Bigfoot—he was eager to follow the giant to the lost gold mines he claimed that Bigfoot guarded!

Wallace eventually moved away from Humboldt County and resumed residence in Toledo, Washington, the area where he and his brothers came from. For the next forty-four years Wallace was on the fringe of the Bigfoot "community," hawking ridiculous facsimiles of Bigfoot footprint casts, bales of "Bigfoot hair," recordings of Bigfoot sounds, and even photos and films of Bigfoot. He had spherical stones supposedly fashioned by Bigfoot for use as missiles to kill game. When Texas oil millionaire Tom Slick turned his attention from the Himalayas to the wilderness around Bluff Creek, Wallace claimed he had captured a Bigfoot, but then could not produce it once a serious offer was made. His outlandish claims and antics were not taken seriously either by the local populace or by the earnest investigators of the mystery. In later years skeptics would point to Wallace as the ultimate source of the Bigfoot phenomenon and would pronounce his marginalization by serious investigators as a sign of shoddy, or even dishonest scholarship.

According to the *Seattle Times* reporter, since Wallace's passing, the surviving family had now come clean, hoping to set the record straight. Wishing to spare the "Bigfoot hunters" further embarrassment, they acknowledged that, while it had all been in good fun, Ray Wallace had indeed created the legend of Bigfoot by donning a simple pair of wooden feet, carved by Rant Mullens, and stomping enormous tracks in the freshly turned earth at the Bluff Creek construction site. They went on to assert that he had also laid down spurious tracks throughout the Pacific Northwest during his travels. Todd McKinley, a great-nephew, told reporters that numerous sets of fake feet were used to plant footprints widely. "They've left tracks all over the Northwest," McKinley said. "I don't know how much they did in Oregon. They've been everywhere with them, and that spanned twenty, thirty years." Reportedly, Great Uncle Ray recruited various nephews to help sow the tracks throughout western forests. What presumably began as a mere joke took on a life of its own; consequently they wanted to set things right. Ray Wallace was Bigfoot, and now Bigfoot was dead—simple as that, according to the Wallaces.

This theory was hardly news. Two decades earlier Rant Mullens himself confessed that he had started the legend of the giant hairy apes of Mt. St. Helens, in the Cascade Mountains of southern Washington, as a young forest ranger with a penchant for practical jokes. He carved his first pair of enormous feet out of slabs of wood in 1928, followed by several more sets. According to Mullens, there were some bootleggers or some kind of outlaw types in the region and the phony tracks were intended to scare them off. Then some friends of his used the carved feet to leave tracks elsewhere in Washington and Oregon that were attributed to Bigfoot. Mullens lost track of the feet until reclaiming them some twenty years later. It was later still that Ray Wallace bought

Rant Mullens displaying a pair of crudely carved feet (Associated Press)

a pair of feet from Mullens and eventually obtained the remaining feet, a disputed transaction that led to an ongoing feud between the two men.

For all his whittling skills, Mullens's carved feet are primitive, to say the least. They are flat and blockish, with squared-off toes and heels. Even Wallace observed, "Those things wouldn't fool anybody. Some of the smartest people in the world study Bigfoot, and Rant's feet are almost like square blocks. I've seen thousands of Bigfoot tracks and they don't look anything like Rant's feet. I've got plaster casts of Bigfoot tracks myself. I've even got the tracks the babies make. They're the real thing. There's no point trying to fool the scientists." And yet the plaster casts that Wallace claims are the "real thing" haven't fooled anybody either. Although they are less blockish than Mullens's, they still are crude stylized attempts to represent a giant foot. Wallace's casts appear stilted and rigid, with no real resemblance to an animated footprint. With minor variations they are stereotyped size-graded clones of one another.

Ray Wallace is connected to all this in only two ways that have been established: First, the men who first reported the tracks to the press were in his employ; and second, the events at Bluff Creek started him on his long career of producing and trying to sell crudely faked track casts and bogus photos, and telling outrageous whoppers about his adventures with Bigfoot, mainly after he had moved back to Toledo, Washington.

Returning to that *Seattle Times* reporter, Young asked me for my reaction to the revelation by the Wallace family that Ray was responsible for the footprints of Bigfoot at Bluff Creek. My response was simple—where are the carved feet? I was by then very familiar with the various tracks from Bluff Creek, including the cast made by Jerry

Crew that the Wallaces were alleging marked the beginning of the Bigfoot prank. I had examined numerous photographs of the original cast and even had a duplicate of the cast in my lab, courtesy of Bob Titmus. I also owned a rare copy of a cast of the same individual made by Roger Patterson in 1964, on Bluff Creek above Notice Creek, which exhibited telling variations from the Crew cast. I had spent several days with John Green, Canadian journalist, author, and longtime associate of Bob Titmus, examining and documenting Titmus's cast collection before its transfer to the Willow Creek Museum after his death. Among these were other casts of the same 16-inch individual from the hamlet of Hyampom in 1963. This most impressive assemblage of original casts resulted from Titmus's investigation of the footprints from Bluff Creek and surrounding environs. Some of these were found in remote locations, well removed from the construction site. I had also examined John Green's photographs of the long lines of tracks along the Blue Creek Mountain Road taken in 1967, perhaps the best-documented incident on record. Don Abbott, archaeologist at the Vancouver Museum, also took important photos of these tracks. If the Wallaces had the carved feet they claimed were responsible for the footprints cast by Crew and Titmus and others at Bluff Creek, it would be an easy exercise to confirm or refute it simply by juxtaposing their fake feet with the footprint casts. Without the feet, their claim was as hollow as any other of Ray Wallace's outlandish yarns.

In addition, I was already familiar with examples of Ray Wallace's casts from news photos of his collections, and from other pictures taken by investigators who had long ago looked into the matter and dismissed it. They were transparent fakes, the product of carved static models that bore no dynamic qualities whatsoever. A characteristic feature that was something of a Wallace "trademark" was a broad, sometimes bulbous forefoot demarcated by an exaggerated imitation of a flexion crease that appeared in some of the original Bluff Creek tracks. In the smaller Wallace casts this crease separates the forefoot from the hindfoot; in the larger casts, this crease splits the ball in two. All of Wallace's casts that I had seen had a very stereotypical appearance and none of them could have been responsible for the 16-inch footprints cast and preserved by Crew.

The *Seattle Times* reporter had not seen the carved wooden feet in question, but he had been assured that they did exist and he was sending a photographer, Dave Rubert, over to shoot them. The resulting photos were very revealing to me, but this had no apparent impact on the reporter's blanket acceptance of the Wallaces' claims. The carved feet were very flat and rough-hewn, with rather sharp edges, especially about the toes. The big toe was simply a broad rectangle, and the triangular little toe of the right foot was not even fully separated from the fourth toe. There were no toe stems

Comparison of the Wallace carved foot (center) with the 17-inch cast (left) made by Jerry Crew at the construction site in 1958 and a 15-inch cast (right) of footprints found by Bob Titmus on a sandbar near Bluff Creek the same year. The carved foot does not match the Crew cast but is a crude facsimile, most likely made in imitation of the Titmus cast. (Courtesy of Rick Noll and Dave Rubert)

evident whatsoever; instead the toe pads were separated from the sole by a wide featureless furrow. The most obvious revelation was that the carved feet were only 15 inches in length and relatively narrow. Crew's cast was of a footprint over 16 inches, nearly 17 inches long and 7 inches wide. It should have been plainly obvious that these carved feet had nothing to do with the documented footprints discovered at the California construction site or any other documented location in the Pacific Northwest.

The spotlighted carved feet do bear some resemblances to the numerous Wallace "casts," such as the bulbous forefoot and pronounced split ball, but also displayed some differing traits. For example, the toe row is arranged along an arc rather than a straight edge. The proportions of the foot are more reasonable. What was the inspiration for the subtle departures from the Wallace signature stereotype?

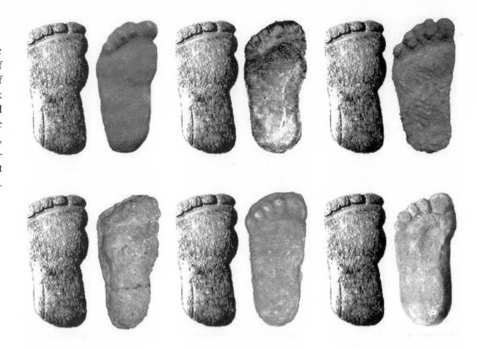

Comparison of the Wallace carved foot (on the left of each pair) to a series of casts from the Bluff Creek region all depicted scaled to same length. Cast in the lower right is from Terrace, British Columbia. The Wallace carving does not match any of the casts. (Courtesy of Bill Miller)

It seems that the carved feet do bear a resemblance to some of the 15-inch footprints that had been found and cast by Bob Titmus on a number of occasions, and also to the well-documented tracks on the Blue Creek Mountain Road, investigated by Green and Dahinden. Titmus had duplicated a pair of deep and distinct 15-inch tracks left in firm sand, and also the larger Jerry Crew cast. In the process he had smoothed them over with clay to facilitate the molding process. This gave the duplicates a slightly blockier and less natural appearance in contrast to the originals I had examined closely. Titmus had distributed a number of copies of these duplicate casts to interested individuals and they were available. Ray's brother, Wilbur "Shorty" Wallace, presented Green with a Titmus duplicate of the Crew cast. He or Ray could very well have had a copy of the Titmus 15-inch cast as well. Could this have been the source of inspiration for the novel carved feet, clearly distinct from the classic Wallace stereotype, only unveiled years after the fact by the later Wallaces?

The resemblance of the carvings to the 15-inch casts is clear but actually fairly superficial. Closer examination reveals the wooden carvings are rather crude copies of the 15-inch casts. John Green has examined and photographed as many of the original 15-inch footprints as anyone. The stretch of tracks along the road on Blue Creek Mountain was laid in fine dust and those numerous prints were exceptionally clear. He noted a number of details that distinguish them from the Wallace's carved wooden feet. First, "being rigid it [the fake foot] cannot change its width when weight comes on it, so it cannot make a clear print wider than itself, but compared to almost all of the 15-inch prints I have casts or pictures of, it is too narrow in relation to its length. That, in spite of the fact that the length of the toes [of the wooden feet] is the shortest of any. Second, it is not sufficiently curvaceous. The living foot can make a print narrower and straighter than itself, but the wooden foot cannot do the opposite. Third, it could not conceivably make a wide straight groove between the toes and the ball [as occurs when the toes are flexed]. Check the picture, which is on the cover of my first printing [*On the Track of the Sasquatch*]. Fourth, the rounding of the heel itself is not symmetrical. All the others are, including the narrow print that shows only four toes, which is otherwise the closest match to the carving."

In fact, the wooden feet don't accurately match *any* of the casts they have been compared to. Coleman points out that Steve Matthes, a member of Tom Slick's Pacific Northwest Expedition, declared 15-inch tracks he found deeply impressed in a sandbar along Bluff Creek in 1960 to be fakes. Coleman asserts these casts to be a match to the Wallace's carved wooden feet and therefore concludes that, "Yes, Wallace appears to have placed prank footprints near some of his California work sites from 1958

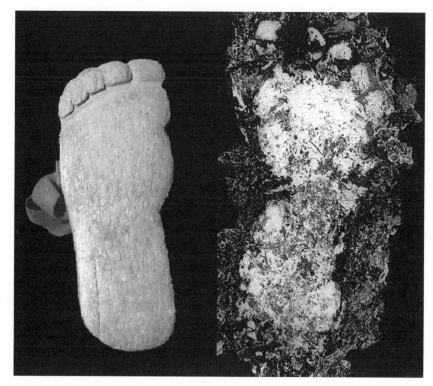

Comparison of the Wallace carved foot (left) and another cast of the 16-inch track discovered at the Bluff Creek construction site. Note the 16-inch foot stepped on and conformed to a stick.

through the 1960s." The problem is—the Matthes cast does *not* match the publicized Wallace carved feet, but the Matthes cast *is* a reasonable match to the Titmus 15-inch casts.

It seems apparent now that the Wallace family was unaware that the pair of carved feet they revealed to the press did not match the original 16-inch footprints cast by Jerry Crew, contrary to their repeated claims. One has simply to look at a picture of the Crew cast alongside the carved foot to realize that the latter could not have produced the former. Furthermore, the imprint of a fair-sized rock protrudes into the original cast along the inside edge of the foot. A rigid wooden foot could not mold around an obstruction like that. It would either high-center the fake foot or be pressed below the level of the imprint. The tracks of this larger individual were observed and cast repeatedly by different individuals and exhibit obvious variations consistent with an animated foot, rather than a rigid fake foot. Additional sets of carved feet are claimed to exist, but my requests to examine and photodocument them were denied.

Another fundamental issue remains—*how* were such crude devices supposedly used to produce hoax footprints convincing to professional trackers and scientists? Could the Wallaces provide a compelling demonstration of how the carved feet were employed to make the tracks that were impressed sometimes an inch or more into firm wet sand? Apparently the media at large felt no obligation to require such a demonstration. The carved feet had simple leather straps attached to their backs and would be worn like primitive snowshoes. However, a snowshoe's ability to reduce pressure on the ground was precisely the effect wearing the enlarged feet would have on the tracks. They would hardly make an impression except in fine dust or wet mud. One reporter for a national news program appeared on screen wearing the fake feet during his story. Standing in wet mud, he attempted to stomp some tracks. To his obvious discomfiture, he could hardly shuffle a step or two as the wet mud sucked onto his disproportionate and stiff fake feet, nearly causing him to pitch over. The mud awkwardly caked onto the wooden feet and the resulting footprints could hardly be distinguished from any other rut or pothole.

Incidentally, examples of the larger 16-inch footprints were examined and cast by Dr. Maurice Tripp, a geologist and geophysicist from San Jose. Tripp's engineering studies of the soil properties and depth of the footprint, which he cast, show the weight of the owner to be more than 800 lbs. In his estimation the tracks were very credible. "It would be difficult to fraudulently prepare hundreds of such tracks overnight—particularly in the type of country in which they were found," he observed.

What about the impressive strides associated with the footprints found at Bluff

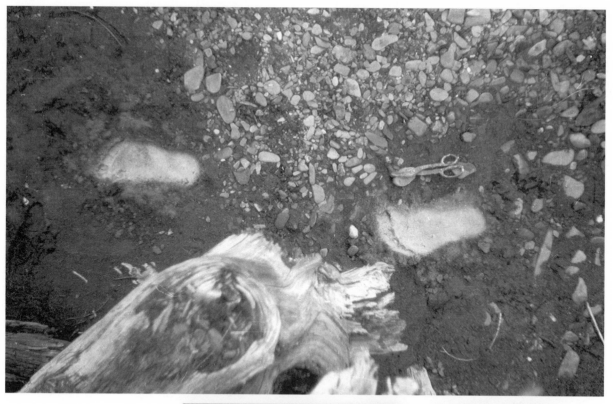

Tracks discovered along Bluff Creek in October 1963 and cast by Al Hodgson (Courtesy of Al Hodgson)

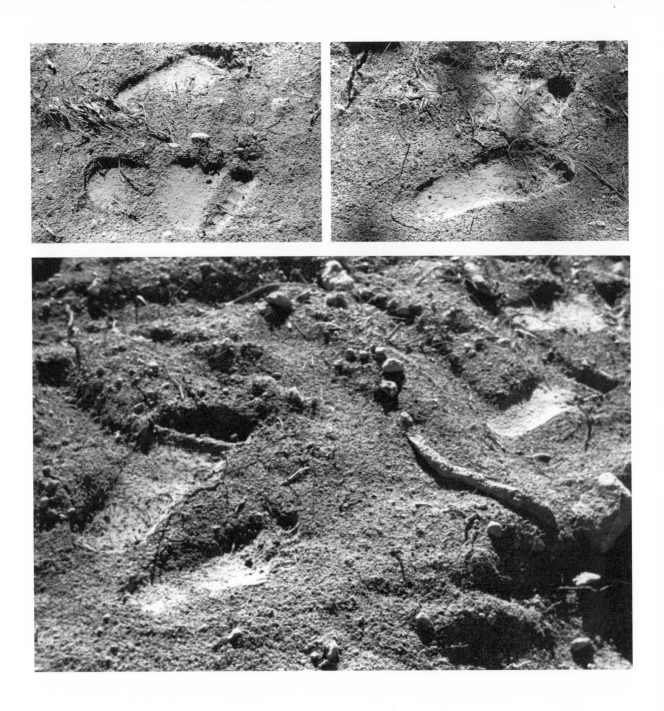

Tracks discovered on the Blue Creek Mountain Road in September 1967 and cast by John Green (Courtesy of John Green)

Creek with a step length of 50–60 inches? The family maintains that the elder Wallace would be pulled along holding onto the tailgate of a slow moving pickup truck in order to accomplish the impressive step length. When challenged to demonstrate this by an exceptionally inquisitive reporter, one of the Wallaces donned the feet and latched onto the tailgate. After a mere step or two, one of the feet came lose and fell off and the Wallace himself nearly tumbled, hanging onto the truck for all he was worth. This escapade was filmed and aired on local television, but obviously didn't receive wide coverage. The conclusion was obvious—it seems the footprint hoaxing enterprise is much easier said than done.

Of course, even if it could be accomplished on a level road, this method could definitely not account for the tracks that marched up and down steep hillsides, presumably in the dark of night. John Green investigated a long line of tracks that generally followed a roadbed while meandering up and down the steep brush-covered banks along the road. The tracks Crew witnessed approached the construction site straight down a steep incline of about 75 degrees. Coming down the incline the track-maker had dug in its heels to purchase more secure footing. Just imagine yourself wearing poorly harnessed snowshoes while traipsing in the dark through heavy brush on steep hillsides. It was rumored that Wallace had somehow rigged logging cables and a spar tree to hoist concrete-weighted feet up and down the hillsides. Never mind that the *original* claim is that the hoaxing was accomplished merely with strap-on wooden feet. The problem is that such a high-lead apparatus requires heavy equipment, including a noisy diesel engine that eats fuel, and inch-thick steel cables. It would likely require at least five men to operate, relying on hand signals that would be impossible to see in the dark. Such equipment is dangerous to operate under the best of conditions, let alone at night. Wallace knew the risks all too well through personal experience. The simple fact remains, there was no high-lead logging operation on the Bluff Creek construction site. Al Hodgson, the proprietor of the hardware store in Willow Creek, spoke to a member of the road crew on Blue Creek Mountain, who said he was the last one to leave at night and the first to return in the morning, and the tracks appeared overnight while he was gone.

So incensed by the cheek of the Wallaces' claim was John Green that he backed the Willow Creek Museum in offering a reward of $100,000 to anyone who could convincingly demonstrate how the Bluff Creek tracks were faked with the Wallaces' carved wooden feet, or by any other method employing means available at that time, nearly half a century ago. Dismissed by some as simply a publicity stunt, the challenge and the monetary reward were and are indeed genuine. Green had personally examined many of the footprints in the ground, including on the nearby Blue Creek Mountain Road and along Bluff Creek itself. Of one occasion he said, "We counted six hundred tracks at Bluff Creek one day in 1967. They showed great variation. The idea that they all could have been made by one carved foot is just nonsense." The challenge stipulates some specific conditions: the tracks must traverse a wide range of terrain, including up and down steep inclines, exhibit variation in shape, toe position, and stride length, sink into ground where human tracks barely leave a mark, be made quietly at night, in areas inaccessible by vehicle—i.e., the conditions of the original Bluff Creek tracks. These conditions are perfectly reasonable to anyone familiar with the primary evidence. To date no one, including the Wallaces, has stepped up to claim the prize.

Ray's son Michael told the *Seattle Times,* "Ray L. Wallace was Bigfoot. The reality is, Bigfoot just died." The article spawned a veritable media frenzy and a demonstration of some of the sloppiest and downright unscrupulous journalism. The story was passed along like a message in that old party game—telegraph. Each successive iteration of the story glossed over some fact, or embellished some unsubstantiated innuendo. Perhaps the most egregious misconduct was by *New York Times* reporter Timothy Egan. Egan at least gave the appearance of doing some serious research into

Casts of the 13-inch (left) and 15-inch (right) tracks on Blue Creek Mountain Road in September 1967

the evidence behind the story, interviewing me, Dr. Henner Fahrenbach, and Dr. Matthew Johnson. After a lengthy telephone interview with me, in which I explained a great deal about the media's misrepresentation of the Wallaces' claims, the straightforward lack of congruence between their carved feet and the original Bluff Creek tracks, and the ridiculous photos and films produced by Ray Wallace, Egan simply rehearsed the same inaccurate story, and, in the process, misrepresented my comments and attributed statements to me that I never made, as well as my colleagues. He went on to label me and any other serious academics involved in the subject as "true believers," even suggesting that we had an "academic investment" in the matter and, therefore, were prone to "scholarly bias." The *New York Times* president, editor in chief, news editor, and editorial page editor were unanimously unresponsive to letters of objection.

A contentious point of the ongoing distortion was what if any role Ray Wallace played in the Patterson-Gimlin film taken in 1967. He claimed to have taken numerous photos and thousands of feet of film of Bigfoot. On at least one occasion he convinced his wife, Elna, to don a gorilla costume so he could film her. "I was Mrs. Bigfoot," she told reporters with a sheepish grin. But rather than show Ray's ridiculously transparent monster costume, the television news repeatedly chose to cut to a clip of the Patterson-Gimlin film to illustrate his claim. To the uninformed the Patterson-Gimlin film became Wallace's film with Mrs. Wallace in the obviously female costume. Mark Chorvinsky played up the association, claiming the Wallaces' admission created profound doubts about the Patterson-Gimlin film. After all, Chorvinsky pointed out, Wallace told Patterson where to go to get a film of Bigfoot. "Ray told me that the Patterson film was a hoax, and he knew who was in the suit," Chorvinsky said. In fact, Wallace asserted it was in the region of Onion Lake (miles from the film site), that it was accessible only by trail (there was actually a logging road close by), that it was near a large boulder (no such landmark). It is clear from what Wallace claimed to have told Patterson about the spot to film Bigfoot that he had no idea where the film was shot. Wallace was a bit more specific in an interview with natural history writer Robert Pyle when he said, "I know exactly which Yakima Indian was in that monkey suit," presumably a reference to Bob Gimlin.

Michael Wallace, Ray's son and family spokesman, told reporters that his father called the Patterson-Gimlin film "a fake," but maintained that Ray had nothing to do with it. Not all Wallace family members were consistent on this point however, and some were confident that Ray was somehow responsible for the incident. Soon the press had so conflated the story that it was being stated outright that the most intriguing photographic evidence for Bigfoot's existence was just another of Wallace's pranks. Eventually, John Hubbell, staff writer for the *San Francisco Chronicle,* attributed to Michael Wallace a comment that the Patterson-Gimlin film "may only be his obliging mother wearing a monkey suit." Stuart Hunter of the *Province* (Vancouver, B.C.) wrote, "Wallace continued the prank for years, culminating in the Patterson-Gimlin 16-mm film footage of an apelike creature walking through a clearing. He told cinematographer-rodeo rider Roger Patterson and Bob Gimlin where to go to spot the creature, which was his wife inside a gorilla suit in the short clip that has withstood independent scrutiny." When the story crossed the Atlantic, the *Telegraph* reported, "The most famous evidence for Bigfoot's existence, the so-called Patterson film, a grainy, cinefilm image of an erect apelike creature, was taken by Roger Patterson, a rodeo rider, in 1967. It was another of Mr. Wallace's fakes, the family said—he told Mr. Patterson where to go to spot the creature and knew who had been inside the suit." The simple acceptance of this unlikely scenario was dramatically demonstrated when I presented a poster at the 2003 meetings of the American Association of Physical Anthropologists on the nature of hominid bipedalism, and illustrated a point with casts of sasquatch tracks and clips from the Patterson-Gimlin film. Repeatedly, my colleagues would interject a comment like "But, I read or heard that the film was shown to be a hoax by some guy with carved wooden feet," or "It was just somebody's wife in a fur suit." Such was the general lack of appreciation of the basic facts of the matter.

I personally contacted Michael Wallace, who was acting as the family spokesperson and who, by the way, wasn't even around in 1958. I was trying to arrange for an opportunity to examine firsthand and properly document the wooden feet that had caused such a commotion. It would presumably be a straightforward exercise to determine which, if any, of the documented Bluff Creek tracks were produced by the Wallaces' carved feet. Wallace was respectful and seemingly sympathetic to my academic request but explained that the family had been quite inundated since the media hysteria and needed time to let the dust settle. He seemed quite sincere in his belief that his father was responsible for hoaxing the original footprints, although it was based merely on an unspoken assumption, since his father never declared during his lifetime to have faked the footprints, other than when he stated that he had tried to throw the hunters off Bigfoot's trail by claiming he had faked the footprints with carved feet. He also distinctly reaffirmed no knowledge of any involvement by his father with the Patterson-Gimlin film and that he had no basis on that account for concluding whether the film was hoaxed or genuine. In the course of the brief conversation, it eventually became clear that a book and possible movie deal were in the works. Subsequently, TLP productions announced they had purchased the rights to the life story of Ray Wallace. "Initially it was just a funny story in the *New York Times,* but the more we learned about Ray, and the ingenious way he captured people's imagination and manipulated the mass media, we knew we had to tell his story," said actor Judge Reinhold. "It's Ray's young son's discovery that his father is Bigfoot, set against the mystery and enchantment of the Northwest woods."

A charming premise to be sure, but shouldn't the world also know what Ray Wallace has said of some of his other "achievements"? Numerous outlandish claims were made in written correspondence to John Green. For example, he claimed to routinely feed a tame Bigfoot apples from his pickup truck window; to have over three hours of film footage of a Bigfoot; that he had been interviewed extensively by government officials who believed the creatures were dropped off by UFOs; to know of lost gold mines that were inhabited by Bigfoots; that Bigfoot skeletons were being sold to Pentagon officials; and that he had captured two Bigfoots, only to have them escape.

It is ironic that Wallace probably had only one successful hoax. He posthumously hoaxed virtually the entire media into believing that he was solely responsible for Bigfoot. He wasn't successful at hoaxing Bigfoot researchers and thus left his family with a lot of worthless artifacts—silly films, hokey casts, and some crude wooden feet. Perhaps the only way his family could get some value from the stuff was by attaching a big story to it. But, if that was the aim, it could only have been realized if the media could be convinced of it, without checking too closely into the basic facts and evidence. Allowing experts to closely examine and document the carved feet or the alleged method of planting the tracks would potentially compromise the story and jeopardize the movie and book deals on the table. Whatever the Wallaces' motivation, the story provides the armchair skeptic with a simplistic explanation for a complex and vexing phenomenon. How is it that the word of a well-known spinner of yarns, if not outright liar, is accepted as gospel, and the accounts of hundreds of credible eyewitnesses who have seen such a primate are dismissed, even when their testimonies are corroborated by footprints, hair, and scat? When it comes to the media's gullibility, it seems that Wallace had the last laugh.

3

WILDMAN OF THE WOODS:
NATIVE AMERICAN TRADITIONAL KNOWLEDGE

Having dealt with the most widely publicized of the "hoaxers," and established some perspective on his impact and its negligible role in the evaluation of sasquatch's potential existence, we return to more primary matters. In the wake of the disproportionate publicity surrounding the events at Bluff Creek during the late 1950s, an influx of accounts of similar discoveries by all manner of outdoorsmen, foresters, and hunters, from over a wide area and extending back for many years prior to 1958, came to light. Since it appeared the local press was actually taking the matter seriously, people came forward and related their own experiences, their fears of ridicule somewhat allayed. A *Humboldt Times* reporter named Betty Allen wondered if the story had even deeper roots. She talked to the local Hoopa and Yurok Indians about their awareness of Bigfoot. One elderly Hoopa man merely reacted: "Good Lord, have the white men finally got around to that?"

To offhandedly dismiss the question of Bigfoot or sasquatch as the outgrowth of some sophomoric prank by a good-humored road builder and his relations is to ignore or trivialize the profound cultural heritage of the earliest inhabitants of this continent. The figure of a hairy wildman occupies a prominent role in those indigenous cultures that inhabit, or have inhabited in the past, areas associated with what is considered by some likely sasquatch range. The Tsimshian mask, already discussed, suggests an early knowledge by Northwest aboriginals of beings with apelike features.

One example of multiple specimens of carved stone heads portraying the *buk'wus* motif, collected in the Columbia River Valley by O. C. Marsh and presently located in the Peabody Museum, Yale University

So do the similarly realistic depictions of ape heads on stone cobbles from the Columbia River.

The Tsimshian people occupy the northern British Columbian coast. The wildman is known in the Tsimshian language as *ba'oosh* or *ba'wes,* which translates to "ape, monkey, Bigfoot, or anything that imitates man." Robert Alley, in *Raincoast Sasquatch,* reports that the Tsimshian "regard the creatures in current folk narratives as just another sort of hominid, albeit hairy and without the benefit of much technology besides rocks and sticks. In the traditional Tsimshian beliefs, he is a member of a forest-dwelling remnant population of hairy 'people' who roam about at the simplest level of existence, but are extremely adapted to a solitary life."

David Rains Wallace, author of *The Klamath Knot: Explorations of Myth and Evolution,* reflects upon the wildman and perhaps approaches the Native American perception of this creature. He ponders the question, "So what wild animal would be hardest for us to discover? An animal very much like us, perhaps. Certain Amazonian tribes were not found until long after their region was 'explored.' Such wild animals couldn't be *too* much like humans. If they were, they would betray their presence by competing for the same habitat. Could an animal be enough like us to escape our endless snooping, yet enough unlike us to escape our endless competitiveness? . . . What if another hominid species had emotionally outgrown *Homo sapiens,* had not evolved the greed, cruelty, vanity, and other 'childishness' that seems to arise with our neotenic nature? What if that animal had come to understand the world well enough that it didn't need to construct a civilization, a cultural sieve through which to strain perception? Such a creature could understand forests in ways we cannot."

The perception of humanness is indeed a difficult "knot" to unravel. Wherever great apes have lived alongside humans, the boundary between the two has been blurred. Symbolism and metaphor have been attached to their unsettling approximation to humanity. Clayton Mack, a First Nations elder of the Bella Coola region of British Columbia, described coming face-to-face with a *boq,* as his people referred to the wildman, through the scope of a rifle. "I look at his lips turning in and out, the top and the bottom too. I look at his face and his chest. The shape of his face is different than the shape of a human being face. Hair over face. Eyes were like us but small. Ears small too. Nose just like us, little bit flatter, that's all. Head kind of looks small compared to the body. Looks friendly doesn't look like he's mad or has anything against us. Didn't snort or make a sound like a grizzly bear . . . I can't; no way am I able to shoot him. I aimed, had my finger on the trigger, pointed it right at the heart. One shot would have killed him dead, just like that. I couldn't shoot him. Like if a person stand over there, I shoot him, same thing. No way I can kill him. My mother told me don't ever shoot a sasquatch. If you shoot them, you gonna lose your wife, your mother or your dad, or else your brother or your sister. It will give you bad luck if you kill them. Leave them walk away."

Along the southern B.C. coast is the Kwakiutl tribe who call the wildman, *buk'wus*. The stylized ceremonial masks depicting the *buk'wus* figure prominently in the works of native carvers. A prominent brow, deep-set eyes, flaring nostrils, a wide grimace exposing large even teeth, and a profusion of hair characterize these masks. An imposing *buk'wus* mask adorns the cover of John Green's book, *Year of the Sasquatch*. I recall happening upon a picture, in a book titled *The First Americans*. It was a photograph of a ceremonial gathering of the Kwakiutl to initiate a chief's son into a secret society. A cast of various animals participated in the drama, represented by crouched actors wearing highly stylized masks and costumes. Prominent in the grouping was an upright figure adorned by a very similar *buk'wus* mask. It impressed me that although the various animals are ritualistically endowed with speech and intelligence and even supernatural powers, the figures nonetheless represent real animals. Why would the *buk'wus* be the singular exception?

The Kwakiutl, like a number of other tribes, distinguish the female element of the wildman separately, as the *dsonoqua,* who is a hair-covered giantess, with large hanging breasts, nocturnal, and fond of abducting children. The *dsonoqua* frequently adorns totem poles and masks, and is distinguished by protruding pursed lips, indicating its whistling call. I recently was told of this alleged behavior of stealing children by a member of a western tribe, who related traditional accounts of crying children being snatched from under the wall of the teepees. Another recounted that her mother often told her and her siblings bedtime stories of Bigfoot, or the *be'a'-nu'mbe'*—the "Brother in the Woods"—in order to settle them down for the night. The mother would

Dsonoqua daughter displaying pursed lips, located in Thunderbird Park on the grounds of the British Columbia Provincial Museum, Victoria, British Columbia. *Dsonoqua* with long outstretched arms, located in Stanly Park, Vancouver, British Columbia. (Courtesy of John Bindernagel)

occasionally tell the children that if they didn't quiet down, Bigfoot would come and reach through the window to snatch them away and no one would know what had become of them. She stressed that Bigfoot was not portrayed as a monster, but rather a long-lost brother, who lives in the mountains and only comes out in times of distress. There was something out there that was large and powerful and should be respected. In a similar vein, a young Native American woman who attended an evening seminar I presented at the Idaho Museum of Natural History remarked afterward that when she was a child, her grandparents had warned her not to venture up certain canyons or the monkey-man would get her.

Although such stories sound incredible and may seem akin to tales of the Bogey-man that serve as an idle reproach to misbehaving children, might they have some basis in fact? A former park ranger in Uganda related an incident to me, in which a chimp stole a native baby that had been parked beside the fields while its mother labored. The infant had been killed and partially eaten before the pursuing villagers could retrieve it. Ethnologist F.W.H. Migeod, while in Sierra Leone, examined a twelve-year-old boy that had been attacked and badly torn by a chimpanzee, and reported this behavior in an historical account from 1926. Under the heading "Man-killing Apes" he wrote, "This species of ape runs to a large size in Sierra Leone" and "noted for its ferocity . . . will without hesitation when it gets the chance attack children and run off with them with the intent to kill them." In a recent news report from Uganda, a growing number of such abductions have come to light. At least eight children have died over a seven-year period and as many were seriously injured. Dr. Michael Gavin, a conservation biologist, who documented one of the most recent incidents observed, "They [chimps] are just trying to get by. If they can't get enough food in the forest, they are going to wander out in search of what's available." It would seem the abduction of human children is a behavior not out of character for an ape or a wildman under the appropriate circumstances.

Adults were not immune to the attentions of the wildman. A storyteller from the Warms Springs tribe in Washington related a traditional account about the fate of tribal hunters. An alarming number of the tribe's men had failed to return from hunting forays. The situation was becoming desperate and the elders asked Coyote to help them determine the fate of their men. Coyote agreed and set out. Soon he encountered Bigfoot who admitted responsibility for the missing hunters. Coyote knew he must put a stop to this or else there would soon be no more Warm Springs people. Coyote challenged Bigfoot to a contest to determine who was more powerful, he or Bigfoot. The loser would refrain from stealing from or killing the men and live a life of seclusion in the mountains. The contest was to close their eyes and vomit up all that they had

The traditional and contemporary theme of abduction of women by sasquatch is portrayed in this caricature (Courtesy of Martin and Erik Dahinden) and finds parallel in the great apes, as depicted in this 1859 sculpture by Fremiet, *Gorilla Carrying Off a Woman.*

eaten that day. The one producing the larger pile was the stronger and thereby the victor. Feeling assured of victory, Bigfoot agreed. Coyote, the trickster that he is, switched the disparate piles while Bigfoot's eyes were closed and promptly declared himself the winner. Surprised by the unexpected outcome, Bigfoot nevertheless accepted the consquences and to this day keeps to himself, secluded in the mountains.

Women may also be the objects of the wildman's special attentions. Alley reports on the rare abduction of women narrated in three tribes that he surveyed. Understandably such incidents are not spoken of freely. He related that, "An elderly Nanaimo [Coast Salish] woman with whom I was acquainted, Mrs. M.B., told me in 1974 that her people had stories, which she believed, of women carried off by *squee'-noos* around the turn of the century. In one case the woman was her grandmother, and either after escaping from the forest, or being returned, she gave birth. The infant was sadly malformed and was in fact stillborn." Once again I was struck by the parallel to very similar accounts emerging from the tribes of the Intermountain West. John Mionczynski, a wildlife biologist in Wyoming was informed of a remarkably similar incident that occurred among the Eastern Shoshone.

Native peoples living in known great ape range likewise have such legends and beliefs. This notion was vividly unveiled to the public in 1859, when French sculptor Emmanuel Fremiet presented *Gorilla Carrying Off a Woman*. It created quite a scandal. Commenting on the reaction, Albert and Jacqueline Ducros observed, "Although the theme of the abduction of a nymph by a faun was a classical one in high art, its avatar, the *Gorilla Carrying Off a Woman* was not found acceptable. Its very *plausibility* made it obscene."

Dr. Herman Rijksen, a Dutch primatologist and orangutan specialist, notes that such legends may well contain a kernel of truth. "The red ape behaves like the classic satyr. On first encounter with a female [human or orang], an attempt at rape is a fully normal way of advertisement for an orangutan male wanting to establish a relationship. . . . Recurrent stories in the Indonesian newspapers of women who claim to have been abducted by orangutan males seem to corroborate the legends. . . . Mawas (orang) males have reliably been reported to go after women to try and wrestle them down for intromission." Recently Biruté Galdikas published a firsthand account of an assault on one of her female field assistants by a male orangutan at a provisioning station. Because of the ape's tremendous strength, Galdikas was unable to dissuade it from culminating the act. Actress Bo Derek experienced an uninvited close encounter with an amorous simian costar while

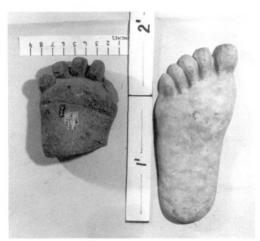

Native American carved stone foot housed in the Centennial Museum at Vancouver, British Columbia, compared to a typical sasquatch footprint cast. The artifact is broken, presumably across the middle of the foot, and the great toe is broken off. (Courtesy of John Green)

filming *Tarzan*. Similarly, actress Julia Roberts, while filming a documentary about orangutan conservation, became the object of a nonconsensual attempted embrace by a large male orang. It required the entire film crew to extricate her from his grip.

Not all Native American cultures have such visible representations as ceremonial masks and totem pole figures. Few have left enduring depictions of their traditional knowledge that are either accessible or recognizable to outsiders. One artifact that I find especially intriguing because of the anatomical insights displayed by the artist is a carved stone foot housed in the Vancouver Museum in British Columbia. John

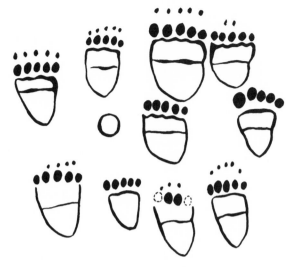

Tracings of Native American petroglyphs depicting bear tracks

Green first drew attention to it and its remarkable resemblances to a sasquatch footprint. The artifact is broken in two places: the great toe, with its larger circumference, is broken off and the heel is broken away. Both the heel and big toe are missing. It seems reasonable, based on the shape and thickness of the carving, that the stone foot would have broken near its middle, precluding some innate weakness or imperfection in the stone. If that is the case, then the reconstructed length and overall dimensions of the carved foot are quite close to those of a typical sasquatch footprint. The anatomical details of note involve the toes. From the undersurface they appear rather short and stubby, and the sole exhibits a distinct crease across the ball and extending laterally to the outside edge of the foot. However, from the top of the foot, the toes appear rather long, such that the crease would fall beneath the "knuckles" of the toes, along the metatarsophalangeal joints. Furthermore, the toes each bear a flat nail, rather than a claw. The broad nails can be seen clearly on end where their edges curl inward. The functional implications of this anatomy will be discussed further in chapter 13. That this artifact exhibits such detail consistent with the anatomy subsequently inferred from numerous contemporary examples of purported sasquatch footprints is quite remarkable.

Kathy Moskowitz, an archaeologist for the Forest Service, has described a unique pictograph at the site of Painted Rock, located on the Tule River Indian Reservation, above Porterville, in the Sierra Nevada foothills of central California. This site is a rock shelter associated with a Native American Yokuts village. The pictographs include a coyote, beaver, bear, frog, caterpillar, centipede, humans, eagle, condor, lizard, and various lines, circles, and other geometric designs, but the most dominant pictograph

Yokuts rock art depicting the Hairy Man (Courtesy of Kathy Moskowitz Strain)

at Painted Rock is that of the Hairy Man, also known as *mayak datat* (*mi!yak datr!atr!*) or *sunsunut* (*shoonshoonootr!*). The Hairy Man pictograph measures 2.6 meters high by 1.9 meters wide, and is red, black, and white. The picture represents an eight-and-a-half-foot high, two-legged creature, with its arms spread out to six-feet wide. It has what appears to be long hair and large, haunting eyes. The Yokuts identify the lines coming from the eyes as tears, because Hairy Man is sad according to their creation story. In this elaborate story each animal has its own ideas about how humans should be created and what qualities they should possess, influenced largely by the behaviors they are known by to the Yokuts. The Hairy Man suggests that humans should walk upright on two legs as he does. But when humans are made they are terrified of the monstrous Hairy Man and flee. This saddens the Hairy Man and so he is depicted with tears streaming from his eyes. As people begin to fill the earth and displace the animals, each animal resigns itself to a restricted niche. Hairy Man said, "I will go live among the big trees (giant sequoias) and hunt only at night when people are asleep."

Moskowitz points out that throughout the story the qualities and natural behaviors of the animals are recounted as they are woven into the tapestry of the story. The Hairy Man appears to be based on a clever animal that is bipedal, nocturnal, and dwells in the forest. A mischievous side is also depicted. The women report the frequent stealing of acorn meal left out to dry. If unattended, it frequently vanished, with only Hairy Man's giant footprints in the sand to attest to his mischief. His habit of striking wood on trees is thought to be in imitation of the pounding of acorns by the women.

A more sinister aspect of Hairy Man appears in the most frequently recounted story still told by the Yokuts today, according to Moskowitz:

BIGFOOT, THE HAIRY MAN

Bigfoot was a creature that was like a great big giant with long, shaggy hair. His long shaggy hair made him look like a big animal. He was good in a way, because he ate the animals that might harm people. He kept the Grizzly Bear, Mountain Lion, Wolf, and other larger animals away. During hot summer nights all the animals would come out together down from the hills to drink out of the Tule River. Bigfoot liked to catch animals down by the river. He would eat them up bones and all.

It was pleasant and cool down by the river on hot summer nights. That is when grownups liked to take a swim. Even though people feared that Bigfoot, the Hairy Man, might come to the river, people still liked to take a swim at night.

Parents always warned their children, "Don't go near the river at night. You may run into Bigfoot."

Now Bigfoot usually eats animals, but parents said, "If he can't find any animals and he is very hungry, he will eat you. Bigfoot, the Hairy Man, doesn't leave a speck or trace. He eats you up bones and all. We won't know where you have gone or what has happened to you."

Some people say Bigfoot, the Hairy Man, still roams around the hills near Tule River. He comes along the trail at night and scares a lot of people. When you hear him you know it is something very big because he makes a big sound, not a little sound.

Children are cautioned not to make fun of his picture on the painted rock or play around that place because he would hear you and come after you.

Parents warned their children, "You are going to meet him on the road if you stay out too late at night." The children have learned always to come home early.

The nature of sasquatchlike beings as understood by the Native American is difficult to appreciate from outside that worldview. Even many contemporary Native Americans themselves have lost touch with the significance of the traditional knowledge of their tribes. In addition, the various tribes have often interpreted their individual experiences in various ways, while still exhibiting many common threads relating to the physical appearance and natural behavior of the sasquatch. Sometimes distinct emphasis is given to particular characteristics. After an enlightening review of the "wildmen" narratives of Northwest tribes, Alley concluded, ". . . it is possible for a dozen different ethnic First Nations to all agree quite closely on the physical attributes of any animal species, perhaps even a number of ascribed behaviors, and yet still hold a belief in any number of different metaphysical attributes for that animal. It is common to hear traditional natives say that while their tribal areas have, for example, a wildman that 'carries a stick and cries like a baby,' that on the other side of the inlet, that other tribal group believes in a wildman that is different because it 'steals women and strikes trees.'"

Gayle Highpine, a Kootenai Indian (the Kootenai tribe's home basically is southeast British Columbia), has traveled extensively among the various reservations and enclaves of North American Indians for the last thirty years. She was a member of AIM, the American Indian Movement, during the '70s. Always interested in the old ways, she has listened attentively to many a medicine man's sasquatch stories as she traveled from reservation to reservation. She has compiled many of these in a rather unique essay, *Traditional Attitudes Toward Bigfoot in Many North American Cultures,* which fills a noticeable void in the anthropological literature, namely a cross-tribal summary of Native American knowledge and beliefs, from the perspective of a Native American. Given this singular accomplishment, the essay is reproduced here in full:

Here in the Northwest, and west of the Rockies generally, Indian people regard Bigfoot with great respect. He is seen as a special kind of being, because of his obvious close relationship with humans. Some elders regard him as standing on the "border" between animal-style consciousness and human-style consciousness, which gives him a special kind of power. It is not that Bigfoot's relationship to humans makes him "superior" to other animals; in Indian culture, unlike Western culture, animals are not regarded as "inferior" to humans but rather as "elder brothers" and "teachers" of humans. But tribal cultures everywhere are based on relationship and kinship; the closer the kinship, the stronger the bond. Many Indian elders in the Northwest refuse to eat bear meat because of the bear's similarity to humans, and Bigfoot is obviously much more similar to humans than is the bear. As beings that blend the "natural knowledge" of animals with something of the distinctive type of consciousness called "intelligence" that humans have, Bigfoot is regarded as a special type of being.

But, special being as he is, I have never heard anyone from a Northwestern tribe suggest that Bigfoot is anything other than a physical being, living in the same physical dimensions as humans and other animals. He eats, he sleeps, he poops, and he cares for his family members. However, among many Indians elsewhere in North America—as widely separated as the Hopi, the Sioux, the Iroquois, and the Northern Athabascan—Bigfoot is seen more as a sort of supernatural or spirit being, whose appearance to humans is always meant to convey some kind of message.

The Lakota, or western Sioux, call Bigfoot *Chiye-tanka* (*Chiha-tanka* in Dakota or Eastern Sioux); *chiye* means "elder brother" and *tanka* means "great" or "big." In English, though, the Sioux usually call him "the Big Man". In his book *In the Spirit of Crazy Horse,* (Viking, 1980), a nonfiction account of the events dramatized by the excellent recent movie *Thunderheart,* author Peter Matthiessen recorded some comments about Bigfoot made by traditional Sioux people and some members of other

Indian nations. Joe Flying By, a Hunkpapa Lakota, told Matthiessen, "I think the Big Man is a kind of husband of Unk-ksa, the earth, who is wise in the way of anything with its own natural wisdom. Sometimes we say that this One is a kind of reptile from the ancient times who can take a big hairy form; I also think he can change into a coyote. Some of the people who saw him did not respect what they were seeing, and they are already gone."

"There is your Big Man standing there, ever waiting, ever present, like the coming of a new day," Oglala Lakota Medicine Man Pete Catches told Matthiessen. "He is both spirit and real being, but he can also glide through the forest, like a moose with big antlers, as though the trees weren't there . . . I know him as my brother . . . I want him to touch me, just a touch, a blessing, something I could bring home to my sons and grandchildren, that I was there, that I approached him, and he touched me."

Ray Owen, son of a Dakota spiritual leader from Prairie Island Reservation in Minnesota, told a reporter from the *Red Wing* (Minnesota) *Republican Eagle*, "They exist in another dimension from us, but can appear in this dimension whenever they have a reason to. See, it's like there are many levels, many dimensions. When our time in this one is finished, we move on to the next, but the Big Man can go between. The Big Man comes from God. He's our big brother, kind of looks out for us. Two years ago, we were going downhill, really self-destructive. We needed a sign to put us back on track, and that's why the Big Man appeared."

Ralph Gray Wolf, a visiting Athapaskan Indian from Alaska, told the reporter, "In our way of beliefs, they make appearances at troubled times," to help troubled Indian communities "get more in tune with Mother Earth," Bigfoot brings "signs or messages that there is a need to change, a need to cleanse," (Minn. news article, "Giant Footprint Signals a Time to Seek Change," July 23, 1988).

Matthiessen reported similar views among the Turtle Mountain Ojibway in North Dakota, that Bigfoot—whom they call *Rugaru*—"appears in symptoms of danger or psychic disruption to the community. When I read this, I wondered if it contradicted my hypothesis that the Ojibways had identified Bigfoot with Windigo, the sinister cannibal-giant of their legends; I had surmised that because I had never heard of any other names for, or references to Bigfoot in Ojibway culture, even though there must have been sightings in woodlands around the Great Lakes, and indeed sightings in that region have been reported by non-Indians. But the Turtle Mountain band is one of the few Ojibway bands to have moved much farther west than most of their nation; and *Rugaru* is not a native Ojibway word. Nor does it come from the languages of neighboring Indian peoples. However, it has a striking sound similarity to the French word for werewolf, *loup-garou,* and there is quite a bit of French influence among the

Turtle Mountain Ojibway. (French-Canadian trappers and missionaries were the first whites that they dealt with extensively, and many tribal members today bear French surnames), so it doesn't seem far-fetched that the Turtle Mountain Ojibway picked up the French name for hairy humanlike being, while at the same time taking on their neighbors positive, reverent, attitude toward Bigfoot. After all, the Plains Cree—even though they retain a memory of their eastern cousins' tradition of the Wetiko (as the Windigo is called in Cree)—have seemed similarly to take on the western tribes' view of Bigfoot as they moved west.

The Hopi elders say that the increasing appearances of Bigfoot are not only a message or warning to the individuals or communities to whom he appears, but to humankind at large. As Matthiessen puts it, they see Bigfoot as "a messenger who appears in evil times as a warning from the Creator that man's disrespect for His sacred instructions has upset the harmony and balance of existence." To the Hopi, the "big hairy man" is just one form that the messenger can take.

The Iroquois (Six Nations Confederacy) of the Northeast—although they live in close proximity to the eastern Algonkian tribes with their Windigo legends—view Bigfoot much in the same way the Hopi do, as a messenger from the Creator trying to warn humans to change their ways or face disaster. However, mentioned among Iroquois much more often than Bigfoot, are the "little people" who are said to inhabit the Adirondack Mountains. I never heard any firsthand stories among the Iroquois about encounters with these "little people"—for that matter, I never heard and firsthand stories in that region about Bigfoot, either—but the Iroquois pass down stories about hunters who occasionally saw small humanlike beings in the Adirondacks (which are not all that far from the Catskills, where Rip Van Winkle was alleged to have met some little bowlers and slept for one hundred years). Some present-day Iroquois assert that the "little people" are still there, just not seen as often because the Iroquois don't spend as much time hunting up in the mountains as they used to. Many Iroquois seem to regard both Bigfoot and the "little people" as spiritual or interdimensional beings who can enter or leave our physical dimension as they please, and choose to whom they present themselves, always for a reason.

Throughout Native North America, Bigfoot is seen as a kind of "brother" to humans. Even among those eastern Algonkian tribes to whom Bigfoot represents the incarnation of the Windigo—the human who is transformed into a cannibalistic monster by tasting human flesh in time of starvation—his fearsomeness comes from his very closeness to humans. The Windigo is the embodiment of the hidden, terrifying temptation within them to turn to eating other humans when no other food is to be had. He was still their "elder brother," but a brother who represented a human potential they feared. As such, the Windigo's appearance was sort of a constant warning to them, a reminder that a community whose members turn to eating each other is doomed much more surely than a community that simply has no food. So the figure of the Windigo is not so far removed from the figure of the "messenger" coming to warn humankind of impending disaster if it doesn't cease its destruction of nature.

The existence of Bigfoot is taken for granted throughout Native North America, and so are his powerful psychic abilities. I can't count the number of times that I have heard elder Indian people say that Bigfoot knows when humans are searching for him and that he chooses when and to whom to make an appearance, and that his psychic powers account for his ability to elude the white man's efforts to capture him or hunt him down. In Indian culture, the entire natural world—the animals, the plants, the rivers, the stars—is seen as a family. And Bigfoot is seen as one of our close relatives, the "great elder brother."

The earliest contemporary recognition of this traditional knowledge of the sasquatch by Native Americans extends back over a century and a half. In 1975, the *Wenatchee Daily World* reported:

Those who think the stories about a huge hairy mystery giant called a Sasquatch are of a recent origin, should talk with Wenatchee Valley College Historian, John Brown. Brown has found evidence that the search for such a legendary creature was underway in the Northwest by the time the earliest white men arrived in the region. While researching material for a book he coauthored with Dr. Robert Ruby—"The Spokane Indians, Children of the Sun"—he came across a passage that must relate to what is now called a Sasquatch.

The reference was in a letter written by the Rev. Elkanah Walker from Fort Colville in 1840. With his wife, Mary, Elkanah Walker was a missionary to the Spokanes. In a letter to the American Board of Commissioners for Foreign Missions, he wrote:

. . . I suppose you will beat with me [sic] if I trouble you with a little of their (the Spokane Indians) superstition, which has recently come to my knowledge.

They believe in the existence of a race of giants which inhabit a certain mountain off to the west of us. This mountain is covered with perpetual snow. They inhabit its top. They may be classed with Goldsmith's nocturnal class, as they cannot see in the daytime. They hunt and do all their work in the night.

They are men stealers. They come to the people's lodges in the night, when the people are asleep, and put them under their skins and take them to their place of abode without their even waking. When they awake in the morning they are wholly lost, not knowing in what direction their home is. The account the Indians give of these *giants* will in some measure correspond with the Bible account of such a race of beings. They say their track is about a foot and a half long. They will carry two or three beams upon their back at once.

They frequently come in the night, steal their salmon from the nets, and eat them raw. If the people are away they always know when they are coming very near by their strong smell, which is most intolerable. It is not uncommon for them to come in the night and give three whistles. Then the stones will begin to hit the houses. The people are troubled with their nocturnal visits.

Brown says he has known about many Spokane Indian legends about monsters but they have been of the Paul Bunyan type that carves out valleys, etc. The ogre referred to in the letter is not really a monster, just a little bigger than man and he had no idea what "mountain to the west is referred to . . . the one that always is snow topped." Perhaps it was Mt. Rainier.

The sasquatch not only occupies a position in the *traditional* folklore of the Native American peoples, it also occurs in *present-day* accounts of encounters or interactions between Native Americans and sasquatch that reiterate the latter's natural flesh-and-blood qualities. To Native Americans, these beings are as much a part of the contemporary landscape as are more familiar animals such as deer or bear, albeit they are more rare and elusive. According to the Native Americans, the sasquatch are elusive because they were created that way and were not intended to be known otherwise, especially not the way the white man "knows" things. If you encounter one, count yourself fortunate, but don't pursue it. Let it be.

Dr. Ed Fusch, anthropologist, has researched the Colville Indians' knowledge of the wildman called *s'cwene'y'ti*. He related a straightforward narrative of a daylight encounter between several Native Americans and a *s'cwene'y'ti* that occurred in June of 1974 on the Columbia River. Sighted at 150 yards distant, the figure was at first taken to be a dark bear standing about eight or nine feet tall, but it turned, walked upright to the river, knelt on his knees with his hands in the water, and put his face in the water for a moment before standing upright again. Two of the witnesses moved closer for a better look and observed the *s'cwene'y'ti* walk in a completely bipedal fashion, stooped slightly forward, with arms swinging. Its forehead sloped back into a dome and the head was set squarely on its shoulders with no visible neck. Two other witnesses whistled, whereupon the *s'cwene'y'ti* stopped, turned its body without moving his feet, and stared at them for a few seconds. Then covering about twenty yards in four or five strides, it stepped up onto a four-foot bank and moved rapidly into the trees.

More recently, Native Americans discovered and photographed large footprints on the mud flats near the Hudson Bay. A sasquatch had apparently raided their nets. Their actions and interest in the incident suggest a desire to comprehend the sasquatch from a contemporary perspective as well as a traditional one.

Although there exists a diverse and sizable literature on the ethnic traditions regarding the wildman persona in Native American cultures, there appears never to have been undertaken a systematic scholarly review and synthesis of the possible universal relationship of these representations specifically with the contemporary notion of sasquatch. This may reflect the frequent and reasonable reticence of the tribal elders to discuss this knowledge openly. The various geographically restricted surveys do present an emerging consensus of a common, but variously interpreted, awareness of the co-existence of an elusive, nocturnal, generally solitary, hair-covered, humanlike animal, inhabiting the mountainous forests of North America. Just how humanlike are these beings in the Native American mind? Clayton Mack concluded, "Half man, half animal, I think. Just like man, but can't make fire, which seems to be all." Unquestionably, these beings occupy a special place among all wild things.

For the Native American as with many ethnic populations around the world, the realms of the "natural" and the "supernatural" exist as one seamless reality. Sasquatch is a definite feature of that reality. The issue for natural science is whether there is in fact a zoological underpinning to the sasquatch traditions, just as there is for the raven or the bear, in spite of the spiritual or mystical embellishments of these characters. One scientist who acknowledged the credibility of this native awareness of the physical existence of sasquatch was Geoffrey Bourne, former Director of the Yerkes Primate Research Center. He said, "That the sasquatch did occur in the northwest of the USA and in British Columbia in Canada in the past is supported by the fact that the Indians of those areas have old legends which tell of creatures like the abominable snowman or sasquatch, tall hairy creatures walking in a bipedal fashion which have been known in that part of the world for generations . . . but the question at issue remains—does a similar creature exist on the North American continent today?"

Dr. Jane Goodall, renowned primatologist and conservationist, spoke openly during an *NPR's Talk of the Nation / Science Friday* (September 27, 2002) interview about her personal conviction that additional species of great ape remain to be discovered, including the sasquatch. She also placed considerable credence in the *contemporary* experiences of Native Americans. Goodall stated, "Well now you'll be amazed when I tell you that I am sure they [sasquatch] exist . . . I have talked to so many Native Americans who've all described the same sounds; two who have seen them." She placed considerable significance on the experiences of these witnesses. Whether sasquatch existed in the past, or still exists in the present, the fundamentally ecological and biogeographical questions remain—What is an ape doing in North America? More fundamentally, how did it get there?

THE GIANT APE OF THE ORIENT:
GIGANTOPITHECUS

"In a clump of bamboo at the edge of a clearing, a troop of giant apes is foraging, snapping huge stalks of bamboo as easily as so many bits of straw. They hear the approach of the men and turn. The two species are eyeball to eyeball for the first time. The apes loom to a height of ten feet—more than twice the height of the men—and weigh well over one thousand pounds . . . one of them lopes forward a few gigantic steps to get a closer look at the group of men. Its head is enormous, with a thickset jaw. The crown of its head looks diminutive atop a huge mandible. Small yellow eyes burn in its skull as it takes in the sight of its furless distant cousin." So Dr. Russell Ciochon, a paleoanthropologist at the University of Iowa, describes a hypothetical en-

counter between *Homo erectus* and *Gigantopithecus* in the forests of eastern Asia sometime within the last one million years. Fossils of both species, recovered from cave deposits in southern China and Vietnam, attest that they overlapped in temporal and geographical range. The nature of their interaction must remain speculative since so little is known about either the full extent of the geographical range or the skeletal anatomy of *Gigantopithecus*.

The giant ape was first discovered when, in the 1930s, paleontologist Ralph von Koenigswald began searching the

Speculative reconstruction of the bust of *Gigantopithecus* (Courtesy of George York)

The silhouettes of a 10-foot- and 8.5-foot-tall bipedal *Gigantopithecus* compared to a 6-foot human figure. Note the tremendous bulk of the giant primates by comparison.

apothecary shops of eastern China for "dragon bones" used as curatives in traditional Chinese medicine. The dragon bones were the fossilized teeth and bones of ancient mammals. In Hong Kong he came upon a single gigantic lower molar tooth of an unknown ape, twice the size of a corresponding gorilla tooth. He named the species *Gigantopithecus blacki*. It was the largest ape that ever lived, estimated to have weighed in excess of 300 kg (660 lbs). By comparison, an adult male gorilla weighs on average 175 kg (380 lbs).

The first massive mandible of this ape was recovered in the late 1950s by an expedition from the Academia Sinica. Two more jaws were to follow, along with hundreds of isolated teeth. In the context of earlier evolutionary interpretations, *Gigantopithecus* was considered to be a hominid, i.e., closely aligned with the human lineage, due to the thick enamel on the rounded molar cusps, its nonprojecting canines, its bicuspid premolars, and the marked posterior divergence of its mandible. This interpretation was a matter of serious discussion up through the 1970s. Alternately, it was suggested that the dental characteristics held in common with hominids demonstrated ecological convergence between *Gigantopithecus* and early hominids. It was hypothesized that these characters, especially thick tooth enamel, evolved in parallel to hominids as adaptations to similar diets and foraging habits. With the fossil remains of *Gigantopithecus* restricted to jaws and teeth, there can only be speculation about the locomotor adaptations of this giant ape. But if the foraging strategies that convergently shaped the jaws and teeth of both *Gigantopithecus* and early hominids produced bipedalism in the one lineage, the possibility of the convergent evolution of bipedalism in *Gigantopithecus* under similar environmental conditions should at least be entertained.

Zoologist and "father" of cryptozoology Bernard Heuvelmans first made a potential link between *Gigantopithecus* and the notorious Himalayan yeti. He suggested, "For giant apes, still agile but no longer able to live in trees, mountains were evidently the most suitable habitat and safest refuge. For the *Gigantopithecus,* which used to live in China, the high Himalayas were the obvious shelter. There, out of reach of their enemies, they could have survived until today, just as their contemporaries have survived in the marshy forests of Borneo and Sumatra. This theory, which is utterly hypothetical, provides the only entirely acceptable explanation of the mystery of the abominable snowman."

When increasing numbers of western mountaineers encountered large bipedal tracks in the Nepalese mountains, the very real possibility of a relic ape was entertained by a number of respected scientists of the day, including Dr. Carleton Coon, University Museum, Pennsylvania State University; Dr. W.C. Osman Hill, Zoological Society of London; and Dr. Adolph Schultz, Anthropological Institute, Zurich. In his 1973 landmark book, *Bigfoot: The Yeti and Sasquatch in Myth and Reality,* Dr. John Napier, former Director of the Primate Biology Program of the Smith-

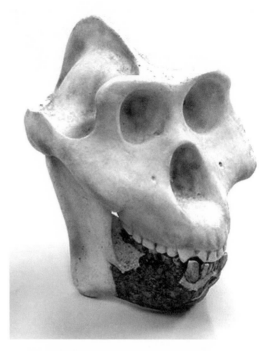

An inferred reconstruction of the skull of *Gigantopithecus blacki,* based on its fossil jaws and teeth, by Grover Krantz

sonian Institution, observed, "It is possible that these creatures, thought by anthropologists to be long extinct, survived in refuge areas such as some of the deep forested river gorges of the Himalayan range until relatively recent times. The absence of a fossil record is not necessarily evidence of extinction."

As recently as 1998, Dr. Chris Stringer of the British Museum of Natural History acknowledged that the yeti legend might not be so far-fetched as often presumed, and may indeed have been inspired by surviving populations of *Gigantopithecus*. He allows that the giant ape may survive today in the dense forests of Southeast Asia, if not in the less likely environs of the Himalayas. Stringer recognized that it would be wrong to assume that yetilike creatures could not survive to the present day without being discovered. "It could have survived until the appearance of modern humans 50,000 years ago, and it is at least possible that it is still living as a very rare creature in remote forest areas," Stringer contemplated. On this matter, Dr. David Begun, paleoprimatologist at the University of Toronto, noted in 2003 that, "There is no reason that such a beast could not persist today. After all, we know from the subfossil record that gorilla-size lemurs lived on the island of Madagascar until they were driven to extinction by humans only 1,000 years ago."

Ciochon has posed a curious paradox. He noted that *Gigantopithecus* is presumably the only great ape to have gone extinct during the Pleistocene. The extant great apes have a greatly reduced range as compared to their former distribution, but they have largely managed to weather the transition to the Holocene. What was it that presumably drove the giant ape to extinction? To account for this apparent inconsistency, Ciochon pointed to the possibility that *Gigantopithecus* had a narrowly specialized diet of bamboo. He then compared the demise of *Gigantopithecus* to that of the giant panda. Thus, dwindling resources, competition, and encroachment by early humans presumably combined to extirpate the giant ape. However, recent analysis of tooth wear seems to have nullified this hypothesis of dietary specialization. The pattern of microscopic pits and scratches on the giant ape's teeth suggests instead that *Gigantopithecus* was a generalized omnivore, removing dwindling resources and competition from pandas as arguable causes of its extinction. The remaining leg of the extinction argument is human encroachment. Other large apes have survived the pressures of human encroachment until now, if only marginally. Why not *Gigantopithecus*?

Gigantopithecus is likely not the only potential ancestral candidate for a relic ape species, however. An extensive radiation of apes is found in the fossil record of the Miocene epoch, beginning about 23 million years ago and ending about 5 million years ago. The Miocene deposits of Africa and Eurasia have yielded an extraordinary diversity of fossil ape species with nearly one hundred extinct species described, and additional species of fossil large-bodied hominoids continue to come to light. For a time the Earth truly was a "planet of the apes." Monkeys had just come onto the scene and begun to diversify as shifting climates created new ecological opportunities for them. Not only were the extinct Miocene apes taxonomically diverse, but also they were quite varied in their adaptations for movement, posture, diet, and distribution. Some were terrestrial quadrupeds, with most of their activity occurring on the ground like modern baboons. Some inhabited temperate woodlands with more seasonal resources and evolved thick tooth enamel to handle a diet of harder coarser foods. A few known species eventually attained great size; a trend real-

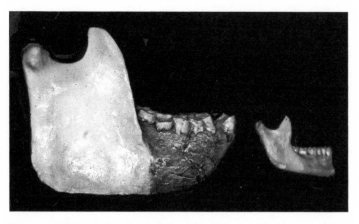

The mandible of *Gigantopithecus blacki* compared to that of a modern human

ized by several species of primate, including a five-foot tall Ice Age ally of the spider monkey found in Brazilian cave deposits, and a gorilla-sized lemur in Madagascar.

An earlier smaller species of the genus *Gigantopithecus, G. giganteus,* is known from the fossil record of Indopakistan dating from 9–6 million years ago. Although the fossil record is patchy, it is believed that *Gigantopithecus blacki,* from the Pleistocene, less than 1.6 million years ago, evolved from its smaller predecessor *G. giganteus,* as the lineage continued to increase in size through the Pliocene. Many species of warm-blooded mammals employed the strategy of gigantothermy to better endure the colder climates of the Pleistocene and to extend their ranges and exploit more temperate habitats.

Of course the prospect of Ice Age relics in the remote corners of Asia is one thing, but what is the likelihood of a giant ape taking refuge in the rugged mountains of North America? A trend for increased body size, thick tooth enamel, and exploitation of terrestrial habitats would conceivably have permitted *Gigantopithecus* to extend its range into more northerly latitudes and better cope with cooler temperatures and more seasonal forest forage spanning across the periodic land bridge between Asia and North America, while its smaller, more arboreal, ape cohorts were restricted to tropical forests.

This trend for increased body size in more northerly-distributed species remains evident even today when comparing related species distributed along a north-south latitudinal cline. Take for example a series of bear species. The spectacled bear (*Tremarctos ornatus*) of tropical South America is the smallest in the series, weighing in at between 60 to 140 kg. Moving northward, we encounter the American black bear (*Ursus americanus*) with weight ranges from 90 to 270 kg. Next comes the grizzly bear (*Ursus arctos*), found throughout the northern Rockies and Canada, weighing between 150 and 380 kg. The Alaskan brown bear (*Ursus middendorffi*), found along the southern coast of Alaska and nearby islands, weighs as much as 675 kg. Finally the polar bear (*Ursus maritimus*) along the northern coastline of Alaska, Canada, and Greenland, can reach weights of 800 kg.

This tendency for increased size in more northerly habitats is referred to as Bergman's rule. It reflects a straightforward strategy for coping with the challenge of retaining body heat in colder environments. Increased body size decreases the ratio of surface area to body mass. The surface area of geometrically similar bodies of different size is the square of a linear dimension, and the volume is the cube of the linear dimension. Take a simple example of a cube with 1-inch sides. Each side has a surface area of 1 square inch (in^2). There are six sides to a cube, so the combined surface area is 6 in^2. The volume is the height × width × depth of the cube, or 1 cubic inch (in^3).

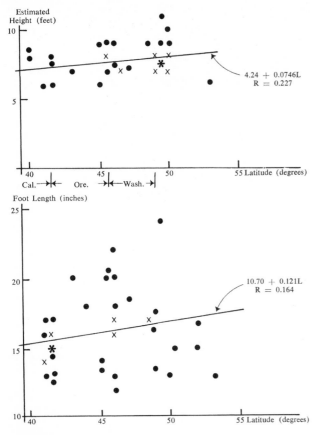

Estimated Height (feet)

4.24 + 0.0746L
R = 0.227

40 Cal. Ore. Wash. 45 50 55 Latitude (degrees)

Foot Length (inches)

10.70 + 0.121L
R = 0.164

40 45 50 55 Latitude (degrees)

An illustration of Bergman's rule. Reported height and foot lengths plotted against latitude. (Courtesy of George Gill)

Now let's double the linear dimensions of the cube making each side 2 inches square, or having a surface area of 4 in². This six-sided cube has a combined surface area of 24 in²—a fourfold increase over the 1-inch cube. The volume of the 2-inch cube increases to 8 in³—an *eightfold* increase. Considering the 1-inch cube, the ratio of surface to volume is 6:1. By comparison, the 2-inch cube has a ratio of 24:8 or 3:1. For larger animals with relatively less skin surface exposed to the environment, less body heat escapes allowing them to tolerate colder climates. Increased body size would be a reasonable adaptation for sasquatch inhabiting northern latitudes, by contrast to its extant tropical cousins.

Bergman's rule also influences body size distribution *within* a single species that has a wide geographic distribution. In 1978, Professor George Gill, a physical anthropologist at the University of Wyoming, presented an analysis of the geographical distribution of footprint lengths and estimated stature of sasquatch. He determined that north-south clines were indeed evident in the data. He concluded, "The present analysis of reports of alleged eyewitness sightings and tracks of the sasquatch was undertaken as the result of the increasing seriousness of this anthropological mystery. The preliminary results of our study support the hypothesis that the sasquatch actually exists, in that population clines in reported body size and track lengths not only seem to exist, but conform to ecogeographical rules."

An added benefit of heat conservation afforded by larger body size is reduced metabolic rate. Less metabolic energy per unit of body mass is required to maintain core temperature. Larger animals also have more space to accommodate a more capacious gut needed to process less easily digested foods. Therefore, larger animals can often subsist on coarser or more seasonal diets of relatively less nutritious foods, as is frequently the condition in temperate forests.

John Green was the first to suggest a link between *Gigantopithecus* and the North American sasquatch. The inferred size and upright posture of *Gigantopithecus* provides ". . . a pretty good thumbnail description of what people have been seeing in North America all along. How it got here is no problem," he said. "Man and a lot of types of animals are believed to have reached North America via a land bridge from Siberia. *Gigantopithecus* . . . could easily have tagged along."

Distinguished scientists such as Geoffrey Bourne, former Director of the Yerkes Primate Research Center, acknowledged the possibility of a large primate immigrating to North America. He suggested, ". . . there is no reason why *Gigantopithecus* could not have earlier come up the mountain causeway and crossed the Bering Strait . . . into the montane forests of America . . . So perhaps the *Gigantopithecus* is the Bigfoot of the American continent . . . Only the discovery of an actual animal and its thorough scientific examination can provide the answer."

Shortly after viewing the controversial Patterson-Gimlin film footage, Dr. Joseph Wraight, chief geographer of the U.S. Coast and Geodetic Survey, shared his reaction to the film with Ivan Sanderson. Wraight echoed Bourne's acknowledgment and countered a frequent misconception about the possible origins of sasquatch: "The presence of large, hairy humanlike creatures in North and Central America, often referred to as sasquatch, appears very logical when the physiographic history of the northern part of this continent is considered. The statement often made that monkey-like creatures never developed in North America may easily be discounted, for these creatures are more humanlike than apelike and they apparently migrated here, rather than representing the product of indigenous evolution. The recent physiographic history of the polar edges of North America reveals that the land migration of these creatures from Asia to America is a distinct and logical possibility. The compelling reason for this distinct possibility is that the land bridge between Asia and North America is known to have existed several times within the last million years, at various intervals during the Pleistocene or Ice Age. It appears then that these hairy, humanlike creatures, sometimes called sasquatch, could easily have migrated to North America at several times during the Ice Age. This is particularly plausible when it is considered that conditions were mild in that area when the land bridges existed. These creatures could have then found conditions along the way similar to their Asian mountain habitat and could naturally have migrated across the land bridge."

Indeed, the temporal and geographical distribution of this giant ape suggests the possibility that it may have spread to North America during times when a land bridge connected Asia and North America. The Bering land bridge is often envisioned as a frozen arctic wasteland or bleak windswept tundra—a seemingly unlikely habitat for an ape. However, the fossil record of past vegetation in the region indicates that during the early to middle Miocene a continuous temperate corridor of deciduous broadleaf and coniferous forest extended from northeast Asia, across the land connection in the region of the present Bering Strait, down across western North America into the Pacific Northwest. Global cooling, beginning about 2.6 mya, first eliminated the broadleaf trees, and then intermittently reduced the coniferous forests, more particularly in the interior regions, while sparing the coastal forests to a greater degree. Of course much of these coastal regions are now under water due to elevated sea levels. Any fossil-bearing sediments from those periods are inundated by hundreds of feet of seawater.

Ongoing discoveries continue to demonstrate the incompleteness of our understanding of movements of species between Asia and North America. A new fossil jaw of a brown bear recovered in southern Canada, dated to 26,000 years ago, has doubled the temporal depth of the fossil record of that species, raising renewed questions about the timing of transcontinental migrations. The discovery in Tennessee of a single molar of the red panda, today found in the Himalayas, with a highly specialized diet of bamboo has obvious geographical, if not ecological implications for discussions of a sasquatch/*Gigantopithecus* connection. This recent find is the second discovery of the red panda molar in North America and dates to between 4.5–7 mya. The earlier find was made in Washington State and dates from 3–4 mya. In a similar vein a new species of grylloblattid, a primitive form of cricket, was discovered in Ape Cave near Mt. St. Helens in the Washington Cascades. The Family Grylloblattidae has the least number of included species and also boasts the most restricted geographical range, including Eastern Asia and northwestern North America. Certainly it would be a worthwhile exercise to compile a list of species of fauna and flora common to these two regions to illustrate the remnant of a contiguous habitat that once spanned these adjacent continents and the extent of their shared communities of plants and animals.

Nearly as controversial as sasquatch itself is the interpretation that a fragment of a fossilized human brow ridge found at Mexico's Lake Chapala may be from the skull of a relic *Homo erectus*. The attribution is a matter of considerable debate, but the close resemblance of the fragment to the cranial anatomy of *Homo erectus* is inescapable. This Asian hominid is thought to have gone extinct within the last 100,000 years, possibly persisting until quite late, until less than 30,000 years ago on isolated Indonesian islands. That the notion of *Homo erectus* in North America is even entertained by serious researchers has implications for the potential range of *Gigantopithecus*. If red pandas, and perhaps *Homo erectus,* both sympatric contemporaries of *Gigantopithecus* in Asia, successfully migrated to North America, what would prevent a similar distribution of *Gigantopithecus?*

Renowned naturalist George Schaller gives serious consideration to the evidence for sasquatch. (Courtesy of Doug Hajicek / Whitewolf Entertainment, Inc.)

Journalist and author John Green is considered the patriarch of proletarian sasquatch investigators. (Courtesy of Doug Hajicek / Whitewolf Entertainment, Inc.)

Native *buk'wus* dancer sculpted by Jack Gibson (Courtesy of Jack Gibson, www.jackgibsongallery.com)

Dsonoqua and her daughter on a totem pole at the Thunderbird Park on the grounds of the British Columbia Provincial Museum, Victoria, British Columbia, Canada (Courtesy of John Bindernagel)

Yokuts rock art depicting the Hairy Man (Courtesy of Kathy Moskowitz)

The mandible of a *Gigantopithecus blacki* alongside Bill Munn's artistic reconstruction of a *Gigantopithecus* bust (Courtesy of Doug Hajicek / Whitewolf Entertainment, Inc.)

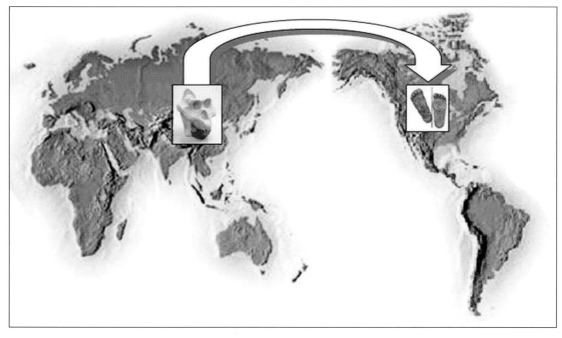

World map depicting the possible expansion of *Gigantopithecus*'s range from Asia to North America during periods when those continents were joined by a land bridge

Rick Noll, longtime field investigator and one of the discoverers of the Skookum imprint (Courtesy of Rick Noll)

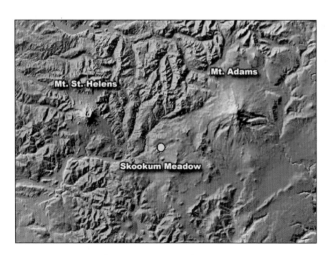

Map indicating the location of the Skookum site in the Southern Washington Cascade Range (Courtesy of Rick Noll)

The Skookum imprint (Courtesy of Rick Noll)

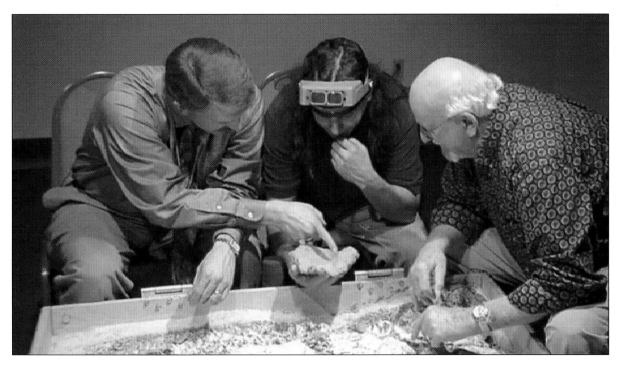

The author, Esteban Sarmiento, and Daris Swindler examine details of a mold taken from the Skookum cast. (Courtesy of Doug Hajicek / Whitewolf Entertainment, Inc.)

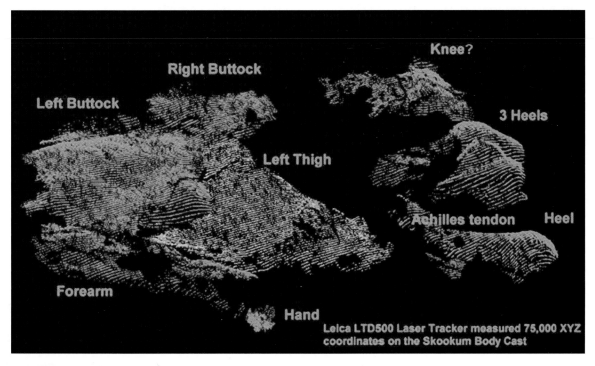

Three-dimensional scan of the Skookum cast highlighting topographic features of the anatomy and their interpretation (Courtesy of Rick Noll)

Sequence of stills from the Freeman video made near Deduct Springs in the Blue Mountains of southeastern Washington (Courtesy of Paul Freeman)

Stills from the Memorial Day video taken by the Pates in northern Washington. Note the increase in height of the subject in the lower still. Inset compares the subject in question (above) to a human model (below) at the same scale. (Courtesy of Doug Hajicek / Whitewolf Entertainment, Inc.)

Forensic team members Doug Divine and Bob Francis prepare to map the Memorial Day video site. (Courtesy of Doug Hajicek / Whitewolf Entertainment, Inc.)

World-class sprinter Derek Prior dons a GPS backpack to track his reenactment of the Memorial Day sighting. (Courtesy of Doug Hajicek / Whitewolf Entertainment, Inc.)

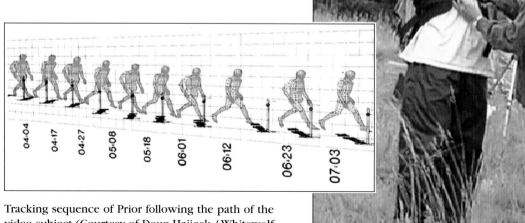

04:04 04:17 04:27 05:08 05:18 06:01 06:12 06:23 07:03

Tracking sequence of Prior following the path of the video subject (Courtesy of Doug Hajicek / Whitewolf Entertainment, Inc.)

Frame 352 of the Patterson-Gimlin film (Courtesy of Martin and Erik Dahinden)

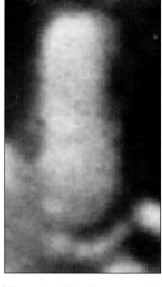

Frame from near the beginning of the film footage showing the back of the subject and the sole of the foot (Courtesy of Martin and Erik Dahinden)

Close-up of the sole of the foot alongside a cast of a footprint from the film site made by Bob Titmus (Note the sole of the foot is viewed at a slight angle.) (Courtesy of Martin and Erik Dahinden)

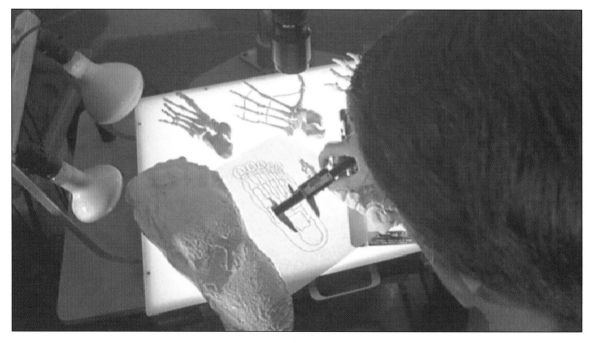

The author correlates dynamic features of a footprint cast from the Patterson-Gimlin film site in order to estimate the skeletal proportions of the sasquatch foot. (Courtesy of Doug Hajicek / Whitewolf Entertainment, Inc.)

Skeletal overlay tracking the Patterson-Gimlin film subject (Courtesy of Doug Hajicek / Whitewolf Entertainment, Inc. and Martin and Erik Dahinden)

Modeling the foot skeleton based on the author's reconstruction of the film subject's skeletal proportions and landmarks (Courtesy of Doug Hajicek / Whitewolf Entertainment, Inc.)

Completed animation of the skeletal kinematics of the film subject constructed by Reuben Steindorf (Courtesy of Doug Hajicek / Whitewolf Entertainment, Inc.)

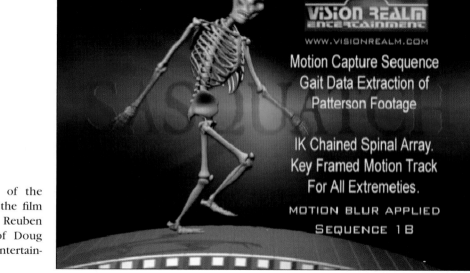

Before and after shots of the apparent quadriceps herniation on the right thigh of the film subject. Inset depicts a close-up of the herniation. (Courtesy of Martin and Erik Dahinden)

The author in his laboratory at Idaho State University discussing the distinctive points of anatomy of sasquatch footprints (Courtesy of Doug Hajicek / Whitewolf Entertainment, Inc.)

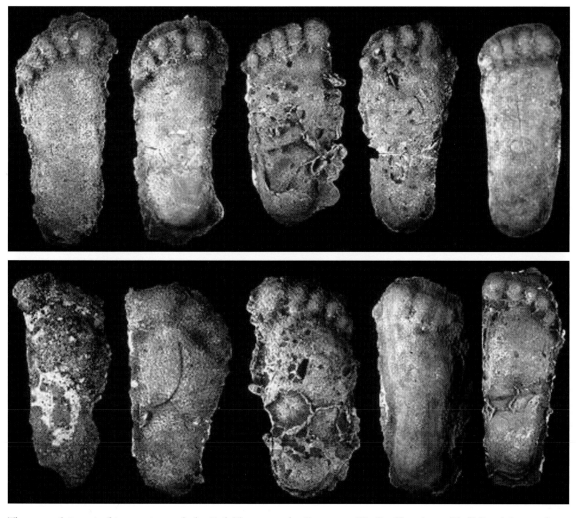

The complete set of ten casts made by Bob Titmus at the Patterson-Gimlin film site at Bluff Creek in northern California, nine days after the filming

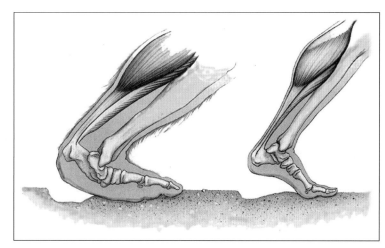

Author's reconstruction of the skeletal basis for the midfoot flexibility that produces the distinctive midfoot pressure ridge in some sasquatch footprints

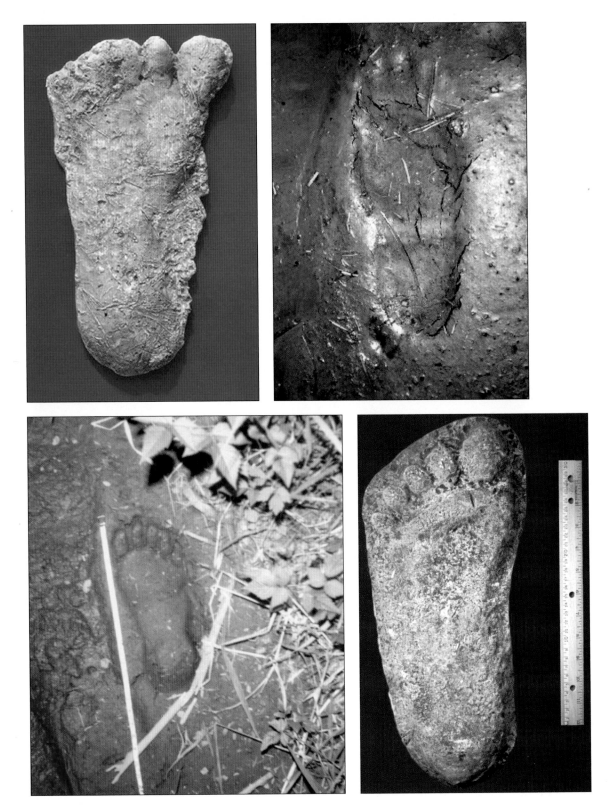

Some examples of typical sasquatch footprints and casts

Latent-fingerprint examiner Jimmy Chilcutt is satisfied that an unknown primate is responsible for the ridge detail on some sasquatch casts. (Courtesy of Doug Hajicek / Whitewolf Entertainment, Inc.)

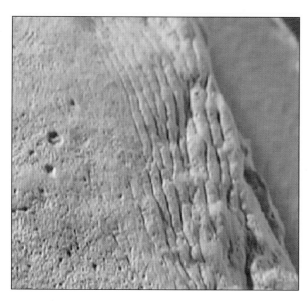

Close-up of ridge detail on the Blue Creek Mountain, California, cast

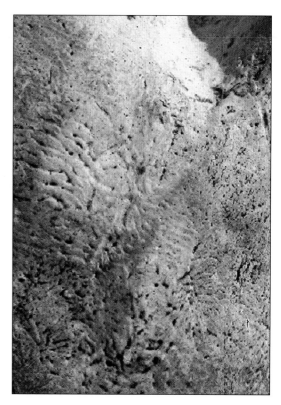

Close-up of the ridge detail and healed scar found on the footprint cast from Walla Walla, Washington

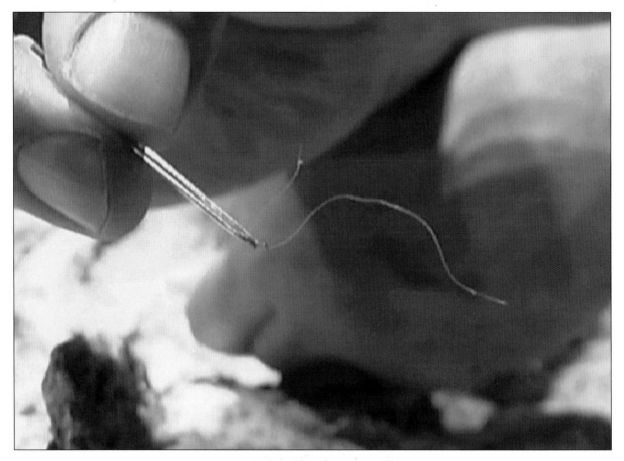

Hair displaying primate characteristics recovered from the Skookum cast (Courtesy of Doug Hajicek / Whitewolf Entertainment, Inc.)

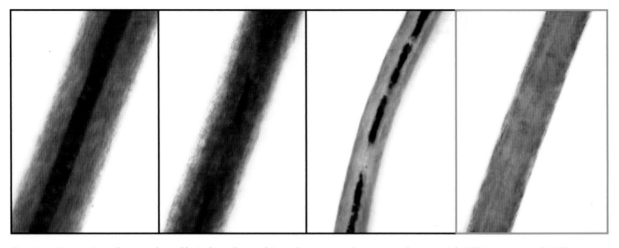

Comparative series of examples of hair from bear, chimp, human, and presumed sasquatch (all to same scale) (Courtesy of Henner Fahrenbach)

The term "migration" sometimes has a misleading connotation in this context. The hypothetical movement of *Gigantopithecus* from Asia to North America should not be considered an intentional trek through unfamiliar and inhospitable territory to invade a new frontier on the American continent. Instead it is more accurately seen as a gradual expansion of a species' range as its habitat expands with changing climate and geography. This expansion may take millennia to culminate, or it may happen more rapidly in species that are naturally more opportunistic and far-ranging. Species that are generalists and more readily exploit diverse resources are also more likely to successfully adapt to varied environments encountered as their range expands. Large body size would further enable them to consume and process a broader repertoire of foods than smaller-bodied apes.

Dr. Tom Agar, a research geologist with the USGS, and expert on the conditions of Beringia, recently speculated, "As for the feasibility of a large primate traveling across northeast Asia, the land bridge, and down into western North America during the Miocene (or later), yes that is possible. And the habitat, climate, edible plants, and coastal marine resources would not have changed all that much along the route during the early to middle Miocene. After that it would have been much continuous coniferous forests along the North Pacific rim, affording somewhat less diverse food resources perhaps . . . Today, the coastal rain forests of southeastern Alaska, British Columbia and parts of the Pacific Northwest are pretty dense, and it might be possible for a critter to avoid detection for a long time."

While most extant primates are generally tropical, this is not an exclusive characteristic of the order. Some living primates, such as the Japanese macaques, *Macaca fuscata,* and the Chinese snub-nosed langurs (golden monki), *Pygathrix (=Rhinopithecus) roxellana,* live today in temperate conditions, and endure very cold, snowy winters. The range of the Japanese macaque is the northernmost known for a nonhuman primate, at 41 30' N., the same latitude, incidentally, as the supposed northern California

Chinese snub-nosed langur (*Pygathrix roxellana*), which inhabits elevations up to 10,000 feet (Courtesy of George Schaller)

habitat of the sasquatch. Furthermore, the Chinese snub-nosed langur, one of the largest monkeys in the world, has adapted to montane mixed bamboo, coniferous, and deciduous forests, up to 3,150 m (10,335 ft) in elevation. They endure the longest and coldest winters of any other primate except humans. The record low temperature in their range is −30°C (−22°F). In the winter their diet consists largely of lichens, a plentiful carbohydrate source, and the inner bark of some trees. A bit to the east, a related species, the black or Yunnan snub-nosed monkey (*Pygathrix bieti*) ranges up to 4,500 m (14,765 ft). Half of its diet consists of lichen from the genus *Usnea,* which hangs from evergreen trees, also found throughout much of the Pacific Northwest.

Apes have a somewhat different physiology and life history than monkeys. In an anonymous peer-review, a member of the scientific board of the California Academy of Sciences argued that the larger brains of apes required a constant and rich food supply during their development—a requirement that could not be met by the resources of the temperate forests of North America. This conclusion ignores the presence of fatty acids critical to brain development in a diet including meat. These fatty acids are especially concentrated in fish, in particular salmon and cold-water species in mountain streams found in the Pacific Northwest and northeast Asia. Furthermore, the reviewer's opinion conveys an oversimplification of great ape evolutionary history, diets, and habitat diversity. For many populations of extant apes living in the tropics, the availability of preferred food resources exhibits marked seasonality. One population of chimpanzees studied, occupied a range of nearly 200 square miles with a mere 3 percent forest cover. Bonobos exhibit a peak in births just after the dry season, presumably to maximize the length of the beneficial rainy season for both lactating mother and infant before the next dry season. Mountain gorillas live in a very cool and wet climate, considering they are found in the tropics, often experiencing subfreezing temperatures at night. The gorillas typically range between the 2,743 m (9,000 ft) and 3,352 m (11,000 ft) elevations, but have been tracked as high as 4,114 m (13,500 ft), the lowest elevation at which snow falls on the Virunga Mountains. They have evolved much thicker coverings of hair than their lowland counterparts. The ability of apes to adopt seasonal life histories may have significance for the potential for sasquatch to adapt to temperate climates.

Furthermore, the remnant species of modern great apes do not fully represent the versatility exhibited by some of their fossil relatives. For example, the ancestral apes became a diverse and successful radiation in Eurasia, where they clearly exploited more seasonal woodland environments. Indeed, much of early ape evolution likely occurred in temperate to subtropical forests in northerly latitudes and possibly higher elevations of Europe and Asia, prior to their expansion into Africa.

Finally, this reviewer's generalization about temperate forests seems to reveal a lack of familiarity with the richness of wild food resources in the rain forests of the west coastal U.S. and Canada available to a large generalized omnivore. In the subalpine habitats of the temperate "rain forests" of the Pacific Northwest and Rocky Mountains, there is an impressive abundance of plant and small mammal resources as I have come to understand through ongoing fieldwork with wildlife biologists and botanists. These potential habitats seem to lie at the core of reported sasquatch activity.

Are there any clues or inferences that address the posture and locomotion of *Gigantopithecus*? Is it reasonable to speculate that bipedalism may have emerged convergently in both early hominids and *Gigantopithecus*? Ciochon continues the search for fossils of *Gigantopithecus*. In his book, *Other Origins: The Search for the Giant Ape in Human Prehistory,* he described the concerted but as yet unsuccessful attempts to locate a skeleton. Without the skeleton of the torso and limbs, the postcranial skeleton, or the base of the skull, which reveals how the skull was poised on the spine, researchers are very limited in what can be inferred about how *Gigantopithecus* stood or walked. The known extant apes are to some extent arboreal and employ powerful arms to suspend themselves below branches where they find the fruit that constitutes a significant fraction of their diet. A common legacy of life in the canopy may have implications for *Gigantopithecus*'s limb proportions. Arboreal apes like the orangutan have proportionately longer upper extremities than do terrestrial quadrupeds such as baboons. Chimps and gorillas spend a great deal of time on the ground, but their apish limb proportions and specialized hip structure constrain them to use a modified form of quadrupedalism called knuckle-walking, although they are capable of walking on only their hindlimbs, when carrying food, nesting material, etc. Jane Goodall observed a chimp in the Gombe, who was stricken with polio leaving its arms paralyzed, which walked exclusively bipedally. The bonobos are even more adept at ambulating bipedally.

Given *Gigantopithecus*'s large size, activity in the treetops was virtually prohibited, and tropical fruits out of reach. It was most likely ground-dwelling, or terrestrial. Once restricted to the ground by its bulk, there would be only two locomotor options—either quadrupedalism (including a possible form of knuckle-walking or perhaps fist-walking) or bipedalism. Dr. Grover Krantz, physical anthropologist at Washington State University, proposed that the configuration of the jaw of *Gigantopithecus* and its relationship to the position of the neck indicate upright posture and bipedalism in this species. Krantz observed, ". . . these [the bodies of the mandible] diverge toward the rear in a remarkable manner. This divergence is so extreme that it would make sense only if the base of the neck was positioned so far forward as to require this spread in

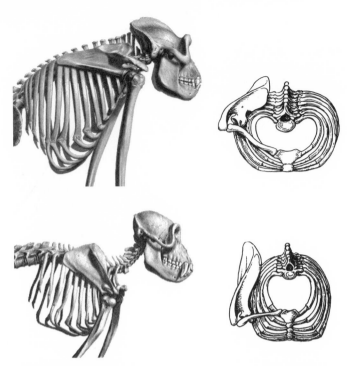

The shape of the thorax and the placement of the shoulder blade in an ape (gorilla) and a monkey (macaque) (Reproduced from a drawing by Jack J. Kunz)

order for it to fit between the jaw's rear extensions. Such a neck orientation would have to be vertical, and thus indicates a fully upright posture . . . With the teeth being of a clear hominoid [ape] design, it follows that *Gigantopithecus* was also a brachiator, like all other hominoids. These animals were much too large for arboreal locomotion, but they still would have had the broad shoulders of that adaptation."

The climbing and arm-hanging behaviors of apes are correlated with a distinctive shape of the chest and shoulder. Apes have a broader flattened rib cage with the shoulder blades positioned on the back of the chest and the shoulder joint directed to the side and slightly upward, whereas quadrupedal monkeys tend to have deep narrow chests, with the shoulder blades positioned on the sides of the rib cage and the shoulder joint pointed downward. The square shoulders of a hominoid are the consequence of a stout clavicle, or collar bone, which provides a bony anchor between the shoulder and the torso, critical in arm-hanging. It also acts as a strut holding the shoulders squarely out from the rib cage. This is very different than the sloping shoulders of an upright bear, for example, with reduced clavicles. These modifications of the ape shoulder render it less adapted for quadrupedal locomotion, since the joint is subjected to shearing forces that would tend to push the arm past the shoulder blade, rather than having the arm and shoulder blade lined up in compression. This would be an even more critical factor in a very large-bodied ape like *Gigantopithecus*. We might expect there to be strong selection to avoid bearing weight on the forelimbs, while favoring bipedalism. Therefore, while Krantz's hypothesis has met with some criticism, it would seem to have a sound basis and hold a better than fifty-fifty chance of being correct—either terrestrial quadruped or terrestrial biped. It certainly deserves consideration pending the discovery of skeletal evidence to refute it.

Living in a mountainous habitat may have also influenced the evolution of bipedalism in *Gigantopithecus*. The mountain gorilla exhibits a trend that may have been paralleled and extended in the sasquatch. Mountain gorillas live on the slopes of the Virunga volcano chain in East Africa. Adults spend over 95 percent of their time on the ground, venturing only occasionally into the trees to feed, and then staying very close to the central trunk of the tree. They do not use their feet in a quadrumanus manner when feeding, as do chimpanzees. Their feeding is usually manual and oral. The foot of the mountain gorilla displays several trends that are associated with its more terrestrial lifestyle. Dr. William Straus describes the distinctions thus: "The foot of the highland gorilla seems to be relatively broader than that of the lowland form; the heel is broader and better developed [longer]; the great toe is relatively longer, and the free digital length much less." By extending these distinctive trends a bit further, one arrives at a basic description of the distinctions of the enigmatic sasquatch footprints. The much greater body size of *Gigantopithecus* and the possibly longer terrestrial tenure in a steep mountainous forest habitat may have shaped its lower extremity into what we now observe as the sasquatch foot.

Cast of the foot of a mountain gorilla

In the past anthropologists had suggested that *Gigantopithecus blacki* was a strict herbivore with a highly specialized diet of bamboo, much like the giant panda. There is evidence of tiny silicate phytoliths embedded in some teeth of *Gigantopithecus,* indicating grasses (bamboo is a grass) in the giant ape's diet, but other evidence from the wear pattern on its teeth indicates a more generalized dietary adaptation. Examination of the pattern of microwear features—pits and scratches—on the surface of the molar teeth shows the greatest similarity between the patterns on the teeth of *Gigantopithecus* and the omnivorous common chimpanzee. So it appears *Gigantopithecus's* diet included not only grasses, but also fruit and seeds of woody plants, insects, and a host of other items. As more is learned about the feeding ecology of the great apes the list of plants eaten continues to grow, in some instances numbering into the hundreds of species. It is also possible that *Gigantopithecus* supplemented their diets with significant amounts of meat, as is the case with chimpanzees. Hence it appears they were generalized omnivores, rather than specialized grazers of bamboo. This could have permitted them to exploit a much broader range of habitats.

Even granting the possibility that the sasquatch may have originated from Asia and may have been in North America for a mere few hundred thousand years, the vexing fact remains that no fossils of such an ape have ever been found on this continent. This time period during which the sasquatch may have inhabited North America is quite brief, geologically speaking. In terms of the fossil record, it is a veritable "blink of the eye," and uncovering fossils from such a particularly short time span would be rather exceptional, especially if the primary habitat were coastal rain forests that are now underwater. Fossilization is actually a rare process, requiring a very specific set of conditions. Most bones are promptly recycled—consumed by scavengers, or carried off by rodents or porcupines—as calcium is a rare commodity in the environment. What bones remain exposed to the elements are rapidly weathered and deteriorated. Should the necessary conditions for fossilization be met, the eventual exposure and discovery of fossils are even further rare and exceptional events.

The recent discovery of a fossil brown bear in Alberta, Canada, illustrates the significant implications of the serendipity of paleontology. Fossil bears were not known from below the extent of the ice sheets prior to 13,000 years ago and all bears south of the glaciers were genetically distinct from those north in Beringia. Where did they come from? The newly discovered fossil turned out to be over 26,000 years old—twice as old as previous southern fossil bears—yet genetically similar to them. "It's always been a mystery why brown bears didn't migrate farther south if they were in Beringia as early as 100,000 years ago, and the passage wasn't blocked until about 23,000 years ago," wondered Paul Matheus, paleontologist at the Alaska Quaternary Center. Although there was an implicit expectation that a population of bears with this distinct genetic identity had extended their range southward much earlier than could be demonstrated in the fossil record at present, there was a complete lack of *any* fossil specimens for a period spanning possibly as much as 80,000 years.

Indeed, the fossil record for most animals is unavoidably spotty. Bob Martin, a paleoprimatologist, estimated there was at the time a total of over 235 known species of living primates, and that 474 extinct species of primate had been described in the scientific literature. Assuming an average species duration of approximately 2.5 million years, based on the number of fossil species known in each stratigraphic interval contrasted with the number known today, Martin estimated that as many as 8,000–9,000 extinct primate species have yet to be discovered in the fossil record. "Our calculations," concluded Martin, "indicate that we have fossil evidence for only about 5 percent of all extinct primates, so it's as if paleontologists have been trying to reconstruct a thousand-piece jigsaw puzzle using just fifty pieces."

In light of such figures, how damning is the lack of fossils for sasquatch? There is an adage among those who search for and study vertebrate fossils—the absence of evidence alone is not the evidence of absence. In other words, just because there is no fossil evidence of sasquatch in North America, does not of itself constitute *conclusive* evidence that an ape species didn't or doesn't exist there. Were it not for a couple of jaw fragments and a number of isolated teeth, there would be no fossil evidence of the past existence of *Gigantopithecus* in eastern Asia. A case in point was the recent discovery of a new species of extinct fossil ape in Thailand. Dubbed *Lufengpithecus chiangmuanensis,* this novel ape, believed to be an ancestor of the orangutan is estimated to have weighed about 70 kg (154 lbs) and to have lived between 10 and 13.5 million years ago. Prior to this initial fossil discovery, the existence of this ape species, its role in its environment, and its possible phyletic relationships would have remained unrecognized. Its recovery and description simply affirm that many more significant fossil discoveries remain to be made.

Considering these factors of fossilization and the incremental pace of new discoveries, the lack of sasquatch fossils in North America is not such a surprising circumstance at all, even allowing that the initial range of sasquatch may have extended beyond the coastal forests that were eventually inundated by rising sea levels. Added to this situation are the especially poor fossilization conditions prevailing in the Pacific Northwest with its moist coniferous forests and predominantly volcanic soils. The acidity of forest soils is not conducive to the preservation of bone.

On the other hand, the acidic groundwater does create caves in limestone sedimentary rocks. The limestone buffers the acidity, allowing exceptional preservation of bone in cave deposits. Where these caves occur they are generally of two types—lateral passages used as denning sites by bear, otter, bats, and small rodents, and vertical sinkholes, some with small openings near the surface. At the bottom of these vertical shafts a cone of detritus accumulates. The in-falling coarse material, wood, and moisture are themselves not conducive to bone preservation. Unless the animals inhabit the caves or are carried there by predators, their remains are not well represented. Sitka black tail deer are common in these regions and yet even their remains are rarely recovered. This biased sampling of the paleofauna is

Articulated skeleton of a giant black bear discovered in a sealed cave on Vancouver Island, British Columbia. Age 10,000 years. (Courtesy of Martin Davis)

being explored and studied by Dr. Timothy Heaton, of the University of South Dakota, and his colleagues, as they excavate the sedimentary deposits in remote caves of southeastern Alaska. The scarcity of the remains of quite common animals is significant; the lack of remains of a much rarer species is therefore not at all surprising, particularly given the infrequent mention of signs of cave habitation present in accumulated sasquatch reports. If the prospect of finding fossil remains of sasquatch in North America is ever realized, they will most likely be fortuitously preserved and discovered in such cave deposits.

Curiously, a scant or altogether absent fossil record for the living great apes has not raised undue concern among paleoprimatologists. Until just recently, there have been virtually no generally accepted fossil ancestors in Africa of chimpanzees or gorillas, despite their presumed presence in Africa for over seven million years. In 2005, Nina Jablonski and Sally McBrearty announced the discovery of a single molar and a pair of incisors identified as the first ever chimpanzee fossils. The three teeth likely came from a single individual that lived about 545,000 years ago. Commenting on this dearth of fossil evidence for living hominoids, Peter Andrews of the British Museum of Natural History said, "The apes seem to have sprung out of nowhere." Obviously this is not literally the case, nor did Andrews intend it to be taken literally as such. Either

Field primatologist Julian Kerbis Peterhans (right) and assistant searching among the litter of the dense Kibale forest for chimp remains (Courtesy of Richard Wrangham and Julian Kerbis Peterhans)

fossil remains of gorillas and chimps, however rare they may be, have yet to be found in African sediments, or they evolved elsewhere and are geologically recent interlopers on the African landscape. In fact, putative fossil ancestors of the African ape/hominid clade *have* been recovered in adjacent Eurasia that presumably migrated into Africa along with other elements of the current African fauna, such as that African symbol, the giraffe. Likewise, only recently has a new species of fossil Asian hominoid been proposed as the most likely ancestor of the orangutan. It bears

The remains of a chimpanzee skeleton partially buried on the floor of the dense forest (Courtesy of Richard Wrangham and Julian Kerbis Peterhans)

striking dental similarities to the orang and originates from the geographic area of Pleistocene orangs. Also of interest, the plants associated with this orang ancestor exhibit a strong and surprising affinity to African flora and indicate a possible, and previously unrecognized, dispersal corridor between Africa and Asia.

Dr. Julian Kerbis Peterhans, of the Field Museum of Natural History, and colleagues have discussed the factors involved in the taphonomy, or the fate of the remains of dead chimpanzees at their site at Kibale, and at other study sites in Africa. In their initial survey they recovered, over a period of nine months, the remains of seven chimps, two that had died quite recently, four that were only represented by cranial remains. Over the next forty-three months only a single chimp was recovered, suggesting that the initial rate of recovery was biased by an accumulation of remains over an extended period prior to collection. At Kibale the pH of the soil is neutral, which contributes to the persistence of the skeletal remains in contrast to the more rapid demise of bone at sites with more acidic soils. Still the significance of finding only a single set of remains in forty-three months at a site of concentrated chimp activity, where multiple researchers are specifically being watchful for such remains, should not be underestimated. Dense forest cover plays a significant role even where conditions are conducive for bone persistence. In nearly 2.3 million km² of forest in Zaire only six archaeozoological sites have been described and all lie at the periphery of the forest with none having produced chimp fossil remains. The paucity of *Gigantopithecus* fossils is better appreciated when considered with such figures in mind. Likewise, the absence of sasquatch fossils should be no more inexplicable.

At the very least *Gigantopithecus* dramatically illustrates the point that one or more species of ape did attain dimensions on the order of those attributed to sasquatch. It is recognized to have lived until quite recently in a region where reports of giant apelike figures that leave oversized footprints are documented today and from whence, in both a biogeographical and paleoecological sense, it could have conceivably extended its range into North American habitats. Although it is certainly tempting to connect the dots, this can only be suggested in a qualified manner, based upon the incomplete evidence at hand—of sasquatch, of the natural history of *Gigantopithecus,* and of the overall biogeographical history of hominoid evolution and dispersal. Nevertheless, the *possibility* is certainly sound and the *plausibility* is quite reasoned and tenable.

Making a Big Impression: The Skookum Body Cast

The fossil record for many species is meager at best, or altogether nonexistent. In some instances only the trace of the organism is preserved—the mold of a shell, the infilled burrow, or the petrified footprints left behind in silt or volcanic ash. Many animals today are reclusive and effectively avoid human contact. Their presence is only betrayed by the subtle tracks and traces they leave behind, commonly during nocturnal activities.

Although infrequently seen for more than a fleeting glimpse, enormous footprints have repeatedly provided mute but intriguing indication of sasquatch's suspected passage. Occasionally impressions of other than oversized humanlike feet are found in the forest—those of hands, knuckles, and buttocks. In one instance, the unmistakable imprint of knuckles was found associated with a short line of 14-inch footprints in the Blue Mountains of southeastern Washington. Had a sasquatch momentarily supported its weight on the backs of its fingers in a decidedly apelike fashion? The cast visibly preserves the impress of the backs of the middle phalanges, or the middle segment of the four fingers, the same parts of the fingers that a gorilla, chimp, or less frequently an orangutan or even gibbon, would rest upon when knuckle-walking, but the sasquatch imprint displays none of the broad thickened

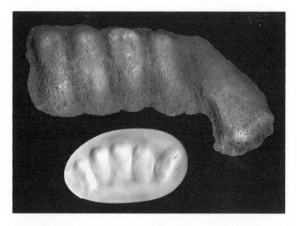

Knuckle prints cast by Paul Freeman, associated with 14-inch tracks, compared to a similar imprint of the author's knuckles made in firm clay. Notice the contrasting orientation of the thumb and the apparent indication of a thumbnail.

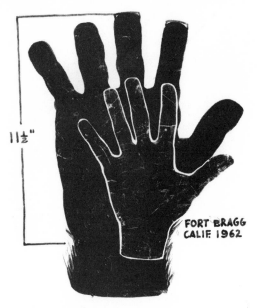

11½"

FORT BRAGG
CALIF. 1962

Outline of the muddy handprint left on the side of a house in Fort Bragg, California, compared to that of a human hand (Courtesy of Chuck Edmonds)

pads of the specialized knuckle-walking fingers of the African apes. The sasquatch knuckles combined measure over 5 inches across, comparable in size to the disproportionately large hand of a male mountain gorilla. What is so impressive about the anatomy of this cast is the subtle, yet readily discerned by the trained eye, trochlear or "spool" shape of the distal ends of proximal phalanges. This shape is especially prominent on the third or middle finger of the cast, it being the largest. On the other digits it is best discerned by lightly running a fingertip over the surface contour of the cast. Despite their imposing breadth, the segments are not exceptionally long and indicate unhumanly short stout fingers. It recalls a muddy 11-inch handprint that was left on the side of a white house near Fort Bragg, California, in 1962. A college art professor, Chuck Edmonds, of Southern Oregon State, traced the handprint. It showed a long palm and proportionately short thick digits, more subequal in length than a human hand. The thumb was relatively long and more in line with the remaining fingers. The muddy print was sufficiently clear that Edmonds could make out details of fingerprints, and observed that no whorls were present.

Returning to the knuckle prints, the end of a very thickly padded thumb juts off from the knuckle imprints, set at what seems to be an odd angle when compared to a human thumb in a hand of similar posture. It would require the thumb to be rotated outward from the remaining fingers, with its pad facing out in nearly the same direction as those of the other fingers, rather than turned toward them, not to mention shorter bones of the palm, the metacarpals. The placement of a distinct impression of the edge of the thumbnail seems to confirm this inferred position. It is this inward rotated set of the human thumb that permits opposition of the thumb to the fingers, critical to the precision grip. The precision grip appears to have been refined relatively late in human evolution and is correlated with the progressive sophistication of tool manufacture.

Interestingly, other handprints, as well as careful eyewitness observations corroborate the apparent lack of opposition in the thumb. When noticed carrying objects or hefting rocks, the sasquatch thumb has been described as lying parallel to the other fingers rather than opposing them. Grover Krantz drew attention to this feature when

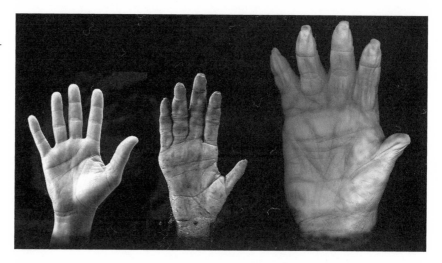

Comparison of the hands of a human, chimpanzee, and gorilla

he reported on two handprints in the *Northwest Anthropological Research Notes* in 1971. The few casts of handprints attributed to sasquatch uniformly exhibit a lack of exceptional enlargement of the thenar muscles, the swelling of muscles at the base of the thumb that lend its drumstick appearance. In humans these muscles are responsible for the distinctive movement of the thumb in opposition, which involves abducting, rotating, and flexing the thumb across the palm. In fact, when these muscles in the human hand become paralyzed and atrophy due to nerve damage, their wasted appearance is referred to as "ape hand." The lack of specialization of the thumb accords with the universal lack of observations of tool use by sasquatch, beyond that exhibited by other great apes, such as brandishing sticks, or lobbing stones.

A handprint found on the bank of a pond and cast by Bob Titmus (Courtesy of Willow Creek–China Flats Museum)

In the 1980s, Bob Titmus had returned to California to continue his search for Bigfoot. He located a line of tracks in a gravel bar that led to a water hole. These were photographed and cast, although the gravelly substrate precluded much detail beyond the contours of the deep impressions. The sasquatch had apparently climbed into the water hole and then climbed back up the steep muddy slope on all fours, leaving a series of slip marks, including one rather clear handprint. The fingers appear elongated due to the slippage, but the most striking feature is the orientation of the thumb imprint. The broad pad of the thumb lies close to the remaining fingers, and its pad is facing in the same direction, rather than rotated toward the palm. Once again, this distinctive anatomy of the nonopposed thumb is evident.

In 1986, Paul Freeman discovered a handprint in rather wet mud. The imprint appears somewhat distorted, presumably as a result of the motion of the hand and the slackness of the mud consistency. The handprint is quite large, over 8 inches across the palm, and correlates with the large 17-inch tracks reportedly associated with it. The hand exhibits the more apelike characteristics of relatively flat palm and nonopposable thumb. It also displays the thick and nearly subequal fingers with extensive webbing between the digits. The flexion creases on the palm and digits are faint, as is to be expected when the hand is bearing weight, but discernable, permitting inferences to be drawn about the placement of joints in the fingers. There is some variability to the arrangement in primary and subsidiary creases, which in humans has given rise to repeated interest throughout history in chiromancy or palmistry. From a strictly functional perspective the primary creases are of two fundamental orientations: running across the palm (transverse) or running the length of the palm (longitudinal). The transverse creases result from the flexion of the knuckles or metacarpophalangeal joints. The transverse creases in Freeman's cast indicate that the palm extends farther past the base of the fingers in a more apelike fashion. Although the fingers appear disproportionately short, this impression is the result of more extensive webbing between the fingers. In humans, the distal transverse crease (the "heart line") typically runs between the second and third digits, while in apes it usually runs completely

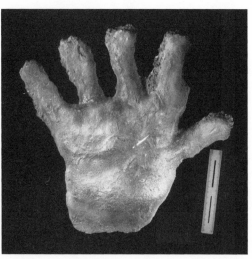

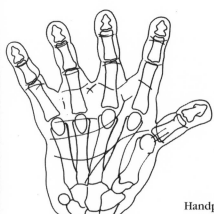

Handprint cast by Paul Freeman and skeletal reconstruction by author based on position of flexion creases

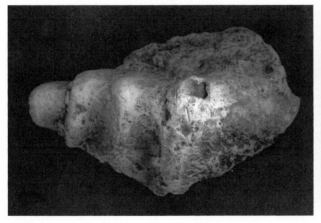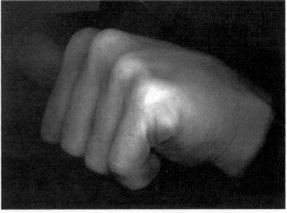

A partial fist imprint cast by Paul Freeman, who at the time mistook it for "toes," compared to the author's hand

across the palm to the gap between the thumb and index finger. This distinction of the human crease is thought to indicate the greater independence of movement of the index finger, while in apes it moves more in concert with the remaining fingers. In this respect the sasquatch handprint more resembles the human configuration. The longitudinal creases result from actions of the thumb. The opposability of the human thumb results in two or more distinct longitudinal creases that extend to the wrist, dividing the "ball" from the "heel" of the palm. These mark the movement of the thumb as it crosses the palm in opposition to the remaining digits. In the sasquatch handprint, these creases are noticeably absent or short, in agreement with the configuration of the nonopposable thumb.

Paul Freeman found another example of a handprint, although he was unable to interpret it. It was found in 1992, north of the Mill Creek watershed associated with a partial footprint of a smaller individual whose tracks had been found repeatedly in the area (discussed later). Freeman had characterized the imprint as "toe marks," but it appears to be the partial imprint of a fist. The outer edge of the palm and the knuckles (metacarpophalangeal joints) of the four fingers are impressed at an angle. What is noteworthy is the relatively subequal size of the knuckles. This detail is consistent with the proportions of the digits evident in other handprints. The impression of the fifth is slightly distorted, but only slightly less prominent than the others. Nor does it appear to be flexed at the carpometacarpal joint, as would be the case in a clenched human fist. Instead the knuckles are more aligned relative to one another. The third knuckle is the most prominent and skin detail in the form of flexion creases are preserved on the third and fourth knuckles.

In May of 1993, east of Walla Walla, Washington, Paul Freeman discovered tracks that led to a sandy bank along Dry Creek. The sasquatch apparently sat down on the

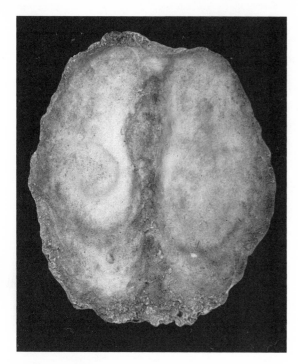

Buttocks imprint found associated with sasquatch tracks and cast by Paul Freeman

edge of an elevated bank and left a distinct imprint of its derrière, which Freeman cast. It measured 14.5 inches across at its widest point. When I first saw the cast, I could not restrain a chuckle (a response I have found the cast evokes from most observers on first exposure to it). Then as I considered it more closely, the appropriate anatomical landmarks became readily apparent. The coccyx or tailbone was evident at the commencement of the deep natal cleft. The cheeks of the buttocks were prominent and apparently well muscled. The depth of the imprint and the clarity and width of the natal cleft indicated a well-developed musculature, with little indication of compressive expansion of excess fat. The delineations of the gluteal folds were evident, separating the upper thighs from the buttocks. The striations of hair could be seen streaming downward and inward across the cheeks toward the natal cleft, but leaving a midline margin of relatively hairless skin; the indications of hair on the backs of the thighs were streaming downward. Below the tailbone was the suggestion of a sphincter and there was no indication of a scrotum, but the possible hint of labia. In spite of my initial involuntary chuckle, this cast was quite "anatomically correct" and yet distinctly nonhuman in dimension and proportion. Its consistency with eyewitness observations of prominent buttocks on sasquatch is noteworthy. The gluteus maximus is a powerful retractor of the hip, important in climbing steep mountainsides, as well as holding the torso erect over the lower limbs. Its apparent development in a giant bipedal ape is appropriate to its posture and habitat.

The singular discovery in September 2000, of a partial body imprint of what appears to be a large hairy animal was something of a sensation. The Bigfoot Field Researchers Organization (BFRO) had mounted a ten-man expedition to the Cascade Mountains of southern Washington. They established camp near a site called Skookum Meadows, an appropriate name, as events would have it. *Skookum* is a Chinook jargon word for "strong" or "powerful," and was often used to refer to evil deities. When used in connection with localities, the word *skookum* generally indicates a place inhabited by a *skookum,* or evil god of the woods. Native Americans avoided *skookum* places, considering them to be haunted. In contradistinction to a *skookum,* a *hehe* was a good spirit and a *hehe chuck* was a fine place for games, races, and other sports and festivities. The geographical distribution of modern equivalents of place names bearing elements of skookum, demon, devil, diablo, or spirit is remarkably coincident with the distribution of contemporary reports of sasquatch encounters or footprint discoveries.

The BFRO expedition's objectives were to detect and document evidence in the field of a large primate, including tracks, hair, scat, signs of foraging, vocalizations, and, if fortunate, a sighting. To this end they set about to experiment with various instruments and attractants. Large broadcasting speakers were set up to test for responses to recordings of not only purported sasquatch calls, but other recordings that might pique a primate's curiosity, such as the sounds of human children playing and infants crying. Thermal imaging and starlight night vision–equipped cameras were on hand. Food baits and pheromone chips were deployed in efforts to lure the quarry to suitable tracking surfaces or, if opportune, into sight. On two separate occasions unidentifiable vocal responses to the broadcasted calls were heard in the distance, once faintly captured on tape. One response call was made from much closer proximity and will be described in chapter 10. Several sets of large, but indistinct, tracks were located and documented. On the morning of Friday, September 22, LeRoy Fish, a professional wildlife ecologist; Derek Randles, a landscape architect and avid outdoorsman; and Richard Noll, an aerospace metrologist, together left camp early to check on the bait that had been laid out during the previous night. At a muddy turnout along a Forest Service road in the general direction from which the previous night's vocalizations had originated, they noticed that a pheromone attractant that had been wired to a nearby tree was missing and several pieces of fruit deposited near a large puddle were gone. Some fruit fragments, bearing what appeared to be broad shovel-shaped tooth marks, were scattered about an unusual depression in the adjacent moist clay soil. The depression appeared to be the imprint of a large hairy animal, as patterns of hair striations covered the various portions of the puzzling configuration. The three investigators considered what sort of animal could possibly be responsible for the impression. One by one, they eliminated the most likely potential candidates—deer, elk, bear, and coyote—and eventually came to the realization that the only other large hairy animal remaining, with the bulk and requisite anatomy, was sasquatch.

The impressions appeared to include that of a left forearm, buttocks, thigh, and heels. But where were the familiar sasquatch footprints? The puddle of water was surrounded by a halo of moist loamy soil. Beyond that perimeter the soil was relatively hard, dry, and choppy from previous traffic—vehicular, pedestrian, and animal. The surface conditions there would not be expected to take the track of a padded foot lacking sharp-edged hooves and claws. Apparently the sasquatch had approached the puddle, lain down across the halo of moist soil on the periphery of the puddle without stepping there, then leaned onto its left elbow and forearm to reach in with its right arm toward the puddle for a sampling of the fruit, while pushing against the mud with its heel. Unfortunately initial references to the site conjured images of a mud wallow, and questions about why a sasquatch would flop into a gooey morass to retrieve some bait. The "mud" was moist soft loamy soil of a consistency ideal for taking up the detailed impression of hair and skin texture without adhering to the body.

Was the sasquatch deliberately trying to avoid leaving tracks in the soft soil? Indications of such behavior have been noted previously in descriptions of sasquatch encounters and are perhaps suggested by the limited tracks left behind. On such occasions the lay of tracks suggested an intentional effort to avoid patches of snow or soft soil that might betray its passing. Would the intentional avoidance of leaving sign be uncharacteristic behavior of an ape, or even beyond its mental capabilities? Recently, Dr. Sue Savage-Rumbaugh, a primatologist from Georgia State University, and colleagues suggested that chimpanzees intentionally break vegetation as a form of trail marker for the benefit of stragglers when the group is spread out during foraging forays. It could be a series of stomped down plants, a broken leafy twig dropped in the trail, or even a twig stuck perpendicularly into the ground. These twigs were pushed at least 8 cm into the ground, clearly as a deliberate act. The "trail signs" were so distinct that if the field researcher arrived at a sleeping tree after the group had already departed, she could readily follow the markers and catch up to the chimps. When traveling a muddy path, however, where their footprints are clearly discernable, the chimps expended no effort to leave trail sign in the thick vegetation, suggesting an awareness of their own footprints as indicators of their passage.

Interpretation of the reclining pose of the sasquatch at Skookum Meadows, based on the imprint (Courtesy of Pete Travers)

I have personally had experience with captive chimps recruited for footprint studies that seemed to indirectly corroborate this awareness of their own footprints that is suggested by Savage-Rumbaugh's observations. The study chimp displayed awareness that he was responsible for the footprints left in a sand track box, and recognition of my interest in examining them. He even took to wiping them away in order to avoid going back into a holding cage while I examined and documented his tracks between bouts. Could the sasquatch be aware that their footprints betray their presence in an area? Anecdotal accounts of the behavior of predators such as bears and big cats would seem to indicate an awareness of their own tracks. It hardly seems unreasonable to infer that a hominoid with a higher level of intelligence would exhibit at least as much, if not more cognition.

There is also the matter of the posture inferred from the configuration of the impression. Is it reasonable to imagine a sasquatch simply taking a timely pause on its nocturnal foraging circuit to recline and munch on a few opportune apples? Given their tendency to hold the trunk more upright (orthograde) as compared to typical quadrupeds, great apes do tend to squat or sit while they are feeding. This can readily be observed at feeding time at any zoological gardens with an ape exhibit. While filming gorillas at the Seattle Zoo, Rick Noll recorded

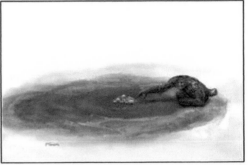
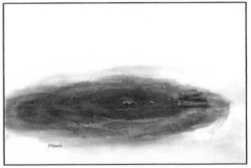

Inferred relationship of the reclining sasquatch to the pond and fruit at the imprint site (Courtesy of Pete Travers)

just such an instance. It was quite intriguing to watch the video of a reclining gorilla resting on its forearm, digging in its heel to adjust its position and reach for apples, taking bites and letting pieces drop to the ground in a manner very reminiscent of the scenes envisioned for the Skookum impression.

Fish, Randles, and Noll, experienced with animal sign, were confident that it was a living animal that made the imprint, but without a clear sasquatch track leading to or from the impression, could they be certain what sort of animal? None of the obvious tracks of hoofed deer and elk, or clawed coyote, present at the scene could be directly associated in a meaningful context with the impression. These tracks simply crossed the imprint, or skirted around it. Of these candidates, the only possible animal with sufficient bulk to create an impression of such dimensions might have been an elk. This would become the focus of our comparisons.

In order to cast the imprint, all the plaster on hand was amassed—over two hundred pounds of it (not such an extraordinary amount when you consider any sasquatch investigator worth his salt has a fifty-pound bag of hydrocal in his trunk). Aluminum tent poles were donated to reinforce the large slab of plaster. The skillfully executed cast, measuring 3½ × 5 feet, was gingerly loaded into a cushioned truck bed and transported back to Derek Randles's workshop. From there, additional scientists were contacted and arrangements were made to clean and examine the cast.

I first learned of the incident just after the cast arrived in Randles's workshop at his home outside of Seattle. I quickly made arrangements to fly there that weekend. I was joined by an initial gathering of scientists that included Fish, who participated in the discovery and casting of the impression; Dr. John Bindernagel, wildlife biologist; Dr. Grover Krantz, physical anthropologist; and Dr. Ron Brown, M.D., an African game guide. Once more, comparisons with impressions of local animals such as elk, deer, bear, and coyote were discussed at length. Could any of these animals be responsible for the shape, dimensions, depth, and attributes of this imprint? The conclusion seemed inescapable—the appearance of the imprint suggested a large hominoid. Preliminary

The author beside the cast of the Skookum imprint, pointing to the most distinctive heel impression (Courtesy of Ron Brown)

Close-up of the heel and Achilles tendon impression on the Skookum cast compared to a cast of a human heel. Note the coyote print in the sasquatch heel impression. (Courtesy of Rick Noll)

measurements indicated the inferred body dimensions to be 40–50 percent greater than my six-foot frame. The unanimous consensus was that this could very well be a body imprint of a sasquatch.

Of particular interest to me was what could only be interpreted as a distinct heel impression. As I meticulously removed the encrusting soil, it appeared that the heel bore skin ridge detail. Once the heel was thoroughly cleaned, a thin latex peel was made of the skin detail. Consultations over the apparent dermatoglyphics, or skin ridges, were had with latent fingerprint examiner, Officer Jimmy Chilcutt. He found them to be consistent in texture and appearance with other specimens of purported sasquatch tracks exhibiting such skin ridge detail (more will be said on this matter in chapter 14).

The impress of the back of the leg above the heel showed a pronounced Achilles tendon, a distinct feature of a bipedal primate. The Achilles tendon appeared relatively broad in concert with the wide heel imprint, and anterior to it was a noticeable hollow that would lie behind the tibia, or shinbone. This pronounced hollow is the consequence of the lengthened heel segment of the foot, an adaptation for increased leverage in response to the massive weight of the sasquatch (discussed further in chapter 13). The imprint of moderate length hair could be seen streaming down across the tendon, giving way to hairless skin just above the margin of the sole, which in turn gave way to the texture of ridged friction skin. From here, the remaining contact surface was painstakingly cleaned of the encasing dirt using dental picks, soft brushes, and

Detail of the surface of the heel impression taken from a thin latex peel of the cast, demonstrating the hair striations coursing across the Achilles tendon, and skin detail over the heel giving way to coarse dermatoglyphics

magnifying lenses. Hairs and fibers were collected in small paper envelopes as they were encountered.

The coincidence of a commercial film crew on hand at the discovery site and the proximity of an adjacent Forest Service road raised some concerns about the credibility of the incident. Could the Skookum cast have been a hoax committed by one or more of the expedition members themselves or perhaps perpetrated upon them by someone else aware of their location and activities? In such a case, the imprint would have to have been laid down at the site sometime between 3:00 A.M. and 6:00 A.M., the interval between the deposition of the fruit and the return of the investigators to the site the following morning. Recall that bites of the fruit were scattered *upon* the imprint (as well as superimposed animal tracks). The hoaxer(s) would have left no sign of their activities while fabricating the detailed impressions of hairy body parts. They would have had to deeply impress, rather than excavate, a coherent shape, as indicated by the flow and extrusion of the loamy soil from beneath a heavy body. They would have had to uniformly apply a rather long hair pattern that was distinct from common animals, such as elk, but suggested similarities to hominoid

Skookum cast with indication of hair flow pattern superimposed and isolated (Courtesy of Owen Caddy)

patterns of hair distribution and flow. This would have been accomplished in an apparently seamless fashion without interruption or over striking of the hair impressions. They would have had to work out the proportions and points of contact of a hominoid body that was about half again as large as a six-foot man. An intentional hoax seemed unlikely, but a case of misidentification remained a possibility in need of further consideration.

Typical posture of a bedding bull elk, with the resulting pattern of impressions. The darkest regions indicate the position of the hooves.

Eventually, careful comparisons to elk imprints were made at multiple game ranches and zoological parks. These comparisons, combined with consultation by professional gamekeepers, ruled out elk as a possible candidate for the imprint. The obvious heel imprints and forearm imprint simply could not be accounted for by the anatomy of an elk. Skeptics opined that the heel imprint was simply the mark of a kneeling elk, without ever examining the cast itself. A wrist of a 650-pound bull elk was obtained by Rick Noll, impressed in soft soil, and cast. Not only did it fail to measure up to the dimensions of the Skookum heel imprint, it was clearly distinct in shape and pattern of hair. The overall orientations of the hair patterns on the Skookum cast were likewise incongruent with those of an elk. This was made quite evident by comparisons to taxidermy museum mounts. And finally, and perhaps most telling, when an elk rises from a repose it must place its hooves directly under its weight in order to stand, leaving tracks in the centerline of its imprint. Yet there are no elk tracks located in the *center* of the Skookum imprint, only deep and clear elk imprints *skirting* the imprint.

A press release announcing the discovery and plans for a coordinated study of the imprint was issued through Idaho State University's office of university relations:

Written by: Glenn Alford, October 23, 2000

ISU Researcher Coordinates Analysis of Body Imprint That May Belong to Sasquatch

Pocatello. Dr. Jeff Meldrum, associate professor of anatomy and anthropology at Idaho State University, is a member of the scientific team examining a plaster cast of what may be the first documented body imprint of a Sasquatch. The imprint of what appears to be a large animal's left forearm, hip, thigh, and heel was discovered Sept. 22 in a muddy wallow near Mt. Adams in southern Washington state by a Bigfoot Field Researchers Organization expedition in the Gifford-Pinchot National Forest. The investigating team, including Meldrum; Dr. Grover Krantz, retired physical anthropologist from Washington State University; Dr. John Bindernagel, Canadian wildlife biologist; John Green, retired Canadian journalist and author; and Dr. Ron Brown, exotic animal handler and health care administrator, all examined the cast and agreed that it cannot be attributed to any commonly known Northwest animal and may represent an unknown primate. Meldrum, whose research includes comparative primate anatomy and the emergence of human walking supervised the careful cleaning of the cast, and will coordinate its analysis by a scientific team. He first became actively interested in the question of the existence of a North American ape after examining fresh Sasquatch (popularly called Bigfoot) tracks in 1996.

"While not definitively proving the existence of a species of North American ape, the cast constitutes significant and compelling new evidence that will hopefully stimulate further serious research and investigation into the presence of these primates in the Northwest mountains and elsewhere," Meldrum said. Dr. LeRoy Fish, a retired wildlife ecologist from Triangles Lake, Ore., with a doctorate in zoology from Washington State University; Derek Randles, a landscape architect from Belfair, Wash.; and Richard Noll, a tooling metrologist from Edmonds, Wash., discovered and cast the partial body imprint. More than 200 pounds of plaster were needed to produce the 3-½ × 5-foot cast of the entire impression, which was reinforced with researchers' aluminum tent poles. Other Sasquatch evidence documented by the party includes voice recordings and indistinct 17-inch footprints. Trace evidence attributed to Sasquatch is usually footprints, but impressions of other body parts, including hands, knuckles, and buttocks, have occasionally been found. This unique instance of a partial body impression provides further insights about this elusive ape species' anatomy. Preliminary measurements indicate its body dimensions are 40 to 50 percent greater than those of a six-foot tall human.

After the cast was cleaned, extensive impressions of hair on the buttock and thigh surfaces and a fringe of longer hair along the forearm were evident. Meldrum identified what appear to be skin ridge patterns on the heel, comparable to fingerprints that are characteristic of primates.

The ridge characteristics are consistent with other examples from Sasquatch footprints Meldrum has studied in collaboration with Officer Jimmy Chilcutt, a latent fingerprint examiner with the Conroe, Texas, Police Department. The anatomy of the heel, ankle, and Achilles tendon are also distinct and consistent with models of the Sasquatch foot derived by Meldrum after examining hundreds of alleged Sasquatch footprints. Hair samples collected at the scene and from the cast itself and examined by Dr. Henner Fahrenbach, a biomedical research scientist from Beaverton, Ore., were primarily of deer, elk, coyote, and bear, as was expected since tracks in the wallow were mostly of those animals. However, based on characteristics matching those of otherwise indeterminate primate hairs collected in association with other Sasquatch sightings, he identified a single distinctly primate hair as "Sasquatch."

Sasquatch is a species of North American ape suspected to inhabit the mountainous forests of the Northwest. Its existence remains controversial despite numerous eyewitness sightings and the discovery of enormous footprints.

The official press release by an accredited academic institution precipitated a great deal of attention and interest in the find, both in the United States and abroad. *New Scientist,* Britain's leading popular science periodical, carried a news feature that portrayed the Skookum impression as the strongest hint yet that Bigfoot or sasquatch may indeed exist. The article also made note of the challenge to get mainstream science to even consider the possibility and examine the evidence. Dr. LeRoy Fish was quoted, "If we can just get other scientists to look at this with an objective view, I think they'll say there must be something out there." Few scientists were willing to take up the challenge. *Field & Stream* magazine ran a news item featuring the cast that precipitated additional responses from a number of outdoorsmen who had encountered inexplicable tracks that resembled those attributed to sasquatch.

Eventually the cast was examined jointly by other expert scientists, including Dr. Daris Swindler, a primate anatomist from the University of Washington and author of *An Atlas of Primate Gross Anatomy,* and Dr. Esteban Sarmiento, a research associate of the American Museum of Natural History with extensive experience studying gorillas in the field. As a fellow anatomist, Dr. Swindler was likewise especially impressed by the appearance of the Achilles tendon. In fact, after a number of past public statements of skepticism concerning the sasquatch phenomenon, the anatomy of the Skookum cast ultimately convinced him of the probable existence of a bipedal North American ape. Dr. Swindler's incisive comments on this feature fully corroborated my preliminary observations: "There is a combination of two muscles, the gastrocnemius and the soleus. In humans about a third of the way down the leg they fuse together to

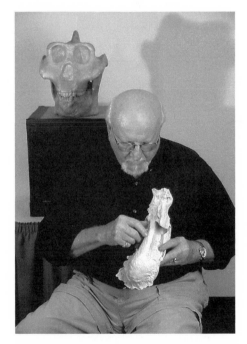

Dr. Daris Swindler, professor emeritus of anatomy, commenting on the implications of the Skookum heel impression (Courtesy of Doug Hajicek / Whitewolf Entertainment, Inc.)

Dr. Esteban Sarmiento, primatologist and research associate of the American Museum of Natural History, was consulted on the analysis of the Skookum cast. (Courtesy of Doug Hajicek / Whitewolf Entertainment, Inc.)

form the tendon of Achilles that then continues on down and attaches to the back end of the heel or calcaneus. *This* (referring to the cast) is a well pronounced, well-developed tendon of Achilles." In nature there is definite correlation between form and function. A well-developed Achilles tendon in a heel impression made by a large ape could be reasonably interpreted only one way. "This was a bipedal animal," concluded Swindler. "In my opinion, the Skookum body cast is that of an upright descendent of *Gigantopithecus*."

The cast reveals another hallmark of bipedalism—a pronounced buttocks, produced by a short flaring hipbone and well-developed gluteal muscles. The gluteus maximus in particular is a powerful retractor of the hip, a particularly important function when walking up steep inclines, characteristic of mountainous terrain. It also aids in keeping the torso upright over the supporting legs when in an erect posture. The presence of the pronounced buttocks in the Skookum cast is correlated with the Achilles tendon, combined indicating an adaptation to bipedalism in this hominoid.

Dr. Sarmiento, with extensive field experience observing primates, was particularly interested in the hair strands, a few which could still be seen trapped in the plaster of the cast. These had been systematically collected during the cleaning of the contact surfaces of the cast, sorted, and identified. As was to be expected, many of the hairs were from such animals as the deer, elk, and coyote, whose tracks were present in the vicinity of the impression. However, a number of the hairs could not be readily identified, although they closely resemble primate hair in many characteristics. These hairs are very similar to an assemblage of independently collected hair samples curated by Dr. Henner Fahrenbach that are suspected of belonging to sasquatch (see chapter 15). Dr. LeRoy Fish independently confirmed that a selection of hairs taken from the cast indeed exhibited primate characteristics.

Dr. George Schaller, renowned naturalist, conservationist, and director of science for the Wildlife Conservation Society in New York City, was afforded an early opportunity to examine the cast. His attention immediately gravitated to the distinct forearm impression and the broad heel imprint. Having studied ungulates, i.e., hoofed mammals, on several continents, Schaller was familiar not only with their anatomy, but also with their behavior. He did not see how the imprint could be attributed to elk bedding. On the other hand, his pioneering field studies of the mountain gorillas made him particularly qualified to draw comparisons to the anatomy of that African hominoid. Although he had frequently witnessed gorillas in the wild reclining during their foraging sessions, he had never taken notice of the impressions they may have left, as the ground was typically heavily vegetated. He summarized his impressions by noting that the Skookum cast, combined with the numerous examples of footprint casts he was familiar with, constituted far more compelling evidence for the existence of the sasquatch than could be mustered for any other prospective unknown hominoid from around the world. He felt a concerted effort to study the matter was certainly warranted.

The Skookum cast has evoked tremendous interest largely due to its novelty—i.e., it's not just another set of footprints. However, it is likely to remain largely a moot point, because its significance is only readily apparent to the highly specialized trained observer. To most onlookers it more resembles a piece of modern art that one might see hanging on the wall of a doctor's waiting room. But for a few careful observers, such as Dr. Swindler, it was the singular piece of evidence that decidedly tipped the scale of his considered opinion.

CAUGHT ON CAMERA: PHOTOGRAPHICS AND FORENSIC MEASUREMENTS

It has been said that a picture speaks a thousand words. It stands to reason that a clear definitive photograph of a sasquatch would go a long way to resolve the question of their existence. In its absence, skeptics point instead to the lack of clear undisputed photography as "evidence" that sasquatch does not exist. Why are there so few unambiguous pictures purporting to show a sasquatch? Of the few claimants, are any credible and compelling? How might the genuine article be differentiated from a clever hoax?

In the age of abundant point-and-shoot cameras and minicamcorders in the hands of a growing number of outdoor recreationists, one might expect an occasional picture, or video recording of sasquatch to materialize. The abundance of quality wildlife documentaries regularly appearing on television has led to the common belief that any animal can be readily located, followed, and filmed in the wild by experienced naturalists and professional cameramen who set out to get the desired footage. I once attended a museum seminar about the conservation of small, less popularized carnivores of the Intermountain West, such as martens, fishers, bobcats, and wolverines. I was particularly keen to learn more about the notorious wolverine, because of its solitary and far-ranging behavior. This predator is largely unfamiliar and poorly understood. The last time I made mention of the wolverine, the only recognition I got from a young listener was a reference to the *X-Men* character. The wolverine is the largest of the weasel family, but more resembles a small bear. A full-grown male may be 4 feet long including tail, stand 17 inches at the shoulder, and weigh 60 pounds. The wolverine is a rare and remarkably elusive, solitary predator that has a very large home range of over 240 square miles. One individual that was radio-collared with a GPS transmitter wandered nearly 550 miles over a six-week period. In one nineteen-day span it traversed a 250-mile stretch from Grand Teton National Park to a ridge just east of my hometown of Pocatello and back again. This individual, at least during his wandering phase, had an estimated home range of 23,000 square miles. The wolverine's preferred habitat range largely corresponds to areas of regional sasquatch reports. The seminar presentation was well illustrated, and I was curious about the protocols used to get such stunning photographs of such a secretive animal as the wolverine. The presenter's response was, "Do you remember that one particular slide with a distant tiny brown dot on a white field of snow? That was the only photo in this presentation of a wolverine taken in the wild. All the rest were simply staged in a pen on a game ranch."

Few people realize that a large portion of the sharp vivid nature photography that adorns calendars and enlivens television features has been staged under controlled conditions. Obtaining truly natural wildlife photographs is a challenging undertaking. With over two years experience in the field with wild chimpanzees, park ranger Owen Caddy related an occasion when a professional wildlife film crew attempted for two months to get film footage of the apes left skittish in the aftermath of a lingering civil war. These expert photographers failed to get any film of the chimps at all, in spite of their concerted efforts to do so.

As surprising as it may sound, no television wildlife production company or wildlife magazine has ever put a professional wildlife photographer in the field for more than a few days in an attempt to obtain photographs, film, or video footage of sasquatch. Production companies that do create programs dealing with sasquatch typically focus their attention on the researchers and theorists (not to mention the skeptics) rather than trying to get original footage. And even if they did make the attempt, the reported behavior of the sasquatch—solitary, nocturnal, furtive, and far-ranging—distinguish it as one of those animals that all wildlife photographers acknowledge as notoriously challenging to even locate and observe, let alone successfully photograph.

Oregon wildlife photographer Michael Durham has captured some of the most stunning images ever of wildlife in their natural setting. He laments the profusion of staged wildlife shots that flood the commercial markets. To his distinction, Durham has developed an innovative system, which he combines with his inherent knack for getting into his subject's head, so to speak, when setting up a location. It takes a great deal of work and he will frequently spend an entire day setting up a single camera trap. He notes that most people have no idea how much work is involved, but that is what it takes to get the quality shot, even of familiar wildlife generally thought to be relatively common in the outdoors.

Others are perhaps less concerned with the aesthetic quality of the shot as they are of capturing an image of some elusive and secretive species of interest. Camera traps are the tool of choice for researcher Jim Sanderson, a research associate with Conservation International's Center for Applied Biodiversity Science. Cameras deployed by Sanderson and other field biologists around the world have documented such rare animals as Abbott's duiker, the Siamese crocodile, and the Andean mountain cat. He remains skeptical of accounts of elusive apes such as the orang pendek. After 250,000 camera trap photos spanning twelve years from one end of Sumatra to the other, not a single photo of orang pendek was taken. While Sanderson doesn't consider North America the primate capital of the present world, he was impressed by photos of footprints emerging from our study areas and what they implied and acknowledged: "A single convincing camera trap photo will blow the lid off it. Go for it."

The typical sighting of a sasquatch is described as quite fleeting, not to mention unexpected. On those infrequent occasions, there is generally only a moment to compose oneself sufficiently and direct the camera at the usually retreating subject. In spite of the odds, a few evocative video clips purporting to show a sasquatch do exist. Analysis of video presents its own set of technical challenges. The image quality of consumer-grade video cameras used by most amateur videographers does not hold up well under stop-action and enlargement. The image quality is susceptible to low-light conditions often encountered in forests, especially at night, and the resulting video usually contains a considerable amount of noise.

An example of a "hit" on a camera trap—this time a black bear (Courtesy of Brian Smith)

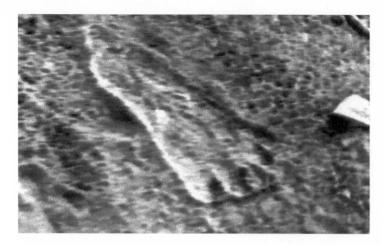

A left footprint from a track at the site of the Freeman video. Notice the long second and third toes. (Courtesy of Paul Freeman)

Paul Freeman, a former Forest Service watershed patrolman in southeastern Washington State, shot a video of a sasquatch in 1994. Even though this particular video was taken in broad daylight from a reasonable distance, it has been afforded very little serious consideration and remains controversial. The footage starts with Freeman shooting a line of fresh footprints that he was informed were found near a spring in the Blue Mountains. The footprints are clearly visible along a dusty trail adjacent to a pond fed by the spring. My interest in the video was initially piqued because of the appearance of these footprints, which measured about 14 inches in

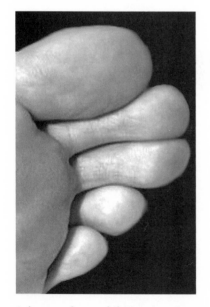

A human foot exhibiting a variant of Morton's foot

length and bore a striking resemblance to footprints I had witnessed myself in the region in 1996. The relative lengths of the toes displayed distinctive proportions. The digital formula was 2=3>1>>4>5, that is, the second and third toes were longer than the great toe, and the last two toes were considerably shorter. This may be considered a variant of what is called the Greek foot, or Morton's foot, in human populations, a condition in which the second toe is longer than the first. It occurs on average less than 10 percent of the time. Having the third toe also longer than the first is an even less frequent variant of the condition. When the first three toes are all of about the same length and the foot is rather broad and squarish, it is referred to as a Peasant foot. The distinction between variants of these respective foot types may be subtle and less than clear-cut. Nevertheless, the fact that the footprints evident on the video exhibited the same distinctive and uncommon toe pattern as was also present in the very compelling and animate footprints I had examined in the same re-

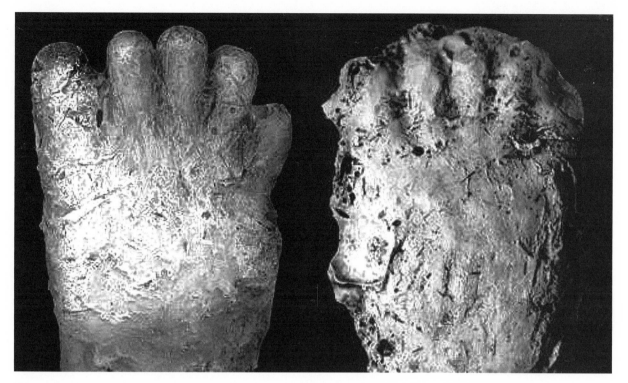

Two examples of footprint casts that exhibit toe lengths that resemble the footprint associated with the Freeman video. On the left is a cast from the region made in 1994, and on the right is a cast made by the author in 1997.

gion just two years later, prompted me to take a more concerted look at the subject of this video.

At the sound of popping brush, the camera rises up from the footprints, and a dark hair-covered figure emerges from the tress and crosses into view. Freeman exclaims "Oh! There he is!" and the subject proceeds deliberately from right to left through the underbrush, passing behind a snag and in front of a small fir, turning its head to glance toward Freeman once before stopping within the partial cover of the brush. From its obscured vantage point it stands motionless and stares back momentarily as if to assess the situation. The camera frame lurches as Freeman attempts to reposition himself for a better perspective, and then again frames the subject as it continues to the left and into the tree line. At no point does the subject appear to be alarmed or threatened by Freeman's presence, merely surprised perhaps that the spring was currently "occupied."

Videographic specialist David Bittner, of Pixel Workshop, examined the footage looking for signs of a hoax. Bittner recounts, "The footage was provided to me on a VHS dub, which I understand was already at least one generation removed from the

David Bittner, professional videographer, at work evaluating the Freeman video (Courtesy of Scott Suchman)

camera original tape. Unfortunately, the original tape was not available for me to analyze. VHS is a relatively low-resolution format, to say the least, and suffers from a significant reduction in detail with each generation of dubbing. The footage was shot with a consumer grade analog camcorder, which was another limiting factor in the quality of the footage.

"We digitized the footage into our video compositing workstation and went to work trying to extract as much usable detail out of the footage as possible. We slowed the footage down, digitally zoomed in, and used a variety of filtering techniques in various combinations. A combination of enlargement, slow motion, and edge enhancement yielded the best results. What remains is, in my opinion, a reasonably compelling video. There is nothing about it technically that points to a hoax, certainly no evidence of editing or digital compositing. There's no sign of a 'zipper on the monkey suit' or anything else that blatantly indicates a hoax, and the gait and apparent mass of the creature fit in with the accepted descriptions of these creatures. Overall, the scene plays out like countless other encounter reports—someone stumbles across a sasquatch; the creature calmly walks away. Freeman's excitement and heightened emotional state are clearly evident in his verbal reactions to what he is witnessing.

"Over the past few years I have viewed numerous photos and videotapes of alleged Bigfoot creatures. I have an interest in the subject, and I never turn down anyone who would like to get my opinion on a video or photo. Some are more interesting than others, of course, and the quality varies tremendously, but typically people tend to overestimate the usefulness of the footage or photos they have collected. A dark blurry image moving through the trees isn't going to capture the interest of any mainstream scientists or legitimate Bigfoot researcher."

Bill Laughery, a retired game warden, and Wes Sumerlin, an outfitter and guide, were familiar with the location of Freeman's video encounter and measured the snag that the subject walked immediately behind. Based on the height of the snag at 11 feet, the subject stood between 7 and 8 feet in height. This agrees with Freeman's statement that the small fir tree situated just behind the subject measured about 16 feet in height. Unfortunately, the Forest Service constructed a campground and other im-

From left to right: Bill Laughery, Ron Brown, "Pee Wee" Sumerlin, and Wes Sumerlin

provements at the spring site, obliterating critical landmarks that were visible in the video. This alteration of the site hindered any subsequent confirmation of scale on which estimates of the video subject's height relied.

During editing of the video footage, producer Doug Hajicek came upon an interesting detail that had apparently otherwise escaped notice, and had never been mentioned by Freeman. In the parting shot, as the subject is seen at some distance from behind retreating into the forest, it appears to bend forward slightly to scoop up something. What seem to be small legs are barely discernable, momentarily dangling before swinging in toward the subject (perhaps securing a hold?). Might this be an infant?

Lori and Owen Pate shot another fascinating piece of video, on Memorial Day in 1996, near Chopaka Lake in northern Washington. The video depicts a dark figure with sure-footed stride, apparent breasts, and possibly an infant on its back, running for cover across a mountainside clearing. One of a number of eyewitnesses present at the time, Tom Lines, viewed it closely, although briefly, through binoculars and noted its covering of hair, which hung long off the arms. He was quite confident it was not a human.

As the location of the Memorial Day video was relatively unaltered since the incident, an elaborate forensic reconstruction was undertaken. Five expert forensic scientists, including Doug Divine of Pacific Survey Supply, and Bill Taft, of Taft & Associates, recreated the incident with the assistance of Derek Prior, three-time all-American sprinter from Washington State University. They set out to determine the size and speed of the figure, and whether a mere human could replicate its performance.

(Left) Owen and Lori Pate reenact their taping of the Memorial Day footage shot in 1996, near Chopaka Lake in northern Washington. (Right) Witness Tom Lines indicates the long hair he observed on the figure he watched through binoculars. (Courtesy of Doug Hajicek / Whitewolf Entertainment, Inc.)

This was perhaps the most high-tech analysis of any film purporting to depict a sasquatch. They effectively mapped the entire hillside recreating the path of the subject and precisely determining the distances it traversed.

After extensive data collection and analysis, the team concluded that the figure was a mere 5' 3" in height, which meant that it was not exceptionally large, and could conceivably have been a human in a costume. But then, should sasquatch exist, they wouldn't spring into being as full grown adults, nor are all adults of the same dimensions. Also, the taller champion sprinter soundly outpaced the figure, although on the first attempt Prior took a potentially nasty spill on the uneven terrain. The team did confirm a curious observation that in the closing sequence the figure gained approximately eight inches in height before passing into the tree line.

Carl Anderson, a forensic measuring expert, directed the analysis conducted at the Memorial Day footage site. (Courtesy of Doug Hajicek / Whitewolf Entertainment, Inc.)

As the team was preoccupied with the scale and speed of the figure, less commentary was offered about the specifics of the anatomy and behavior that could be observed from the video image itself. The subject reportedly hesitated at the tree line for a considerable time and was seen occasionally peering from behind some cover. When it finally broke from cover and was captured on videotape, it made its descent down the hillside, apparently with something of a slightly lighter hue on its back. That something then appeared to slip lower on the figure's back and an appendage of sorts, flailed behind. At that point the running figure, with arms pumping, reached back with its right arm to clasp the object to itself, without breaking stride. Also discernable were unmistakable breasts that gyrated with each running step. There are no indications of clothing, no bulging, bunching, or borders. There is nothing to contradict Lines's testimony that he saw a hair-covered figure through the binoculars. The figure passed behind an intervening rise in topography and when it emerged beyond, its right arm was poised upward, toward or above its head, as if steadying something. The arm dropped and the figure noticeably increased in height, before entering the timber. Could this be a female sasquatch, perhaps a young female, with a youngster on board? The smaller size and apparent breasts suggest this might be the case. Was the infant nearly dislodged from its piggyback position as its mother made a dash for cover behind the intervening high ground?

Infant apes frequently ride atop their mother's shoulders, which may account for the increase in height of the Memorial Day video subject.

Did it receive a boost into a position behind the female's neck and peer over the top of her head? This is a posture for infant transport that is typically practiced by apes, and humans for that matter. Infants ride atop the adult's back or shoulders and peer over the adult's head. This sometimes continues until the infant is quite large, if the adult is indulgent.

Dr. Esteban Sarmiento's impression of the Memorial Day footage was that the behavior of the subject was quite natural—running for cover. But the shortcomings of the medium and the distance from the subject come to the forefront again. "If I had to reach a conclusion, I would have to sit on the fence given the resolution. It's very difficult to prove with certainty that the creature is real or a man in a monkey suit. The detail just isn't there," concluded Sarmiento.

By far the most intriguing and most discussed, albeit controversial, piece of photographic evidence is a 16mm film taken in northern California in 1967, by Yakima, Washington, rodeo rider Roger Patterson and horse rancher Bob Gimlin. Patterson had become interested in the subject after reading an article by zoologist Ivan Sanderson on the subject of "California's abominable snowman." He had made a trip to the region to see the enigmatic footprints for himself, and made a cast of a 17-inch track in

(Left) Frame 352 of the Patterson-Gimlin film taken in October 1967 in northern California. (Right) Close-up of the film subject. (Courtesy of Martin and Erik Dahinden)

1964 on Bluff Creek above Notice Creek. In 1967, word of fresh footprints reached Patterson by way of Al Hodgson, a local proprietor in Willow Creek. Patterson was producing a documentary in hopes of financing a full-time search for the animal. He wished to get footage of clear footprints to include in the production. He enlisted the help of Bob Gimlin, who had accompanied him on earlier trips to Mt. St. Helens, Washington, and the two began patrolling the area surrounding Bluff Creek. For nearly two weeks they found no sign. Then on October 20 they rounded a bend in the creek, their approach and view obscured until the last moment by a large root wad. Suddenly the horses became spooked by a large hair-covered figure about sixty feet to their left, which turned and with a determined, but unhurried gait began to retreat upstream across a wide elevated sandbar. Roger grabbed the camera from his saddlebag, a maneuver he had actually practiced extensively in anticipation of just such an incident, and ran across the creek in pursuit. At the high sandbar he stumbled and dropped to his knees. The subject pivoted at the waist, swinging its arms widely, as it seemed to glare back at the intruders. Roger stayed down momentarily and steadied the camera. What he saw through the viewfinder, Gimlin saw even more clearly from atop his horse—an upright figure covered with short, dark hair that was slightly longer on its head. Its body was thick and heavily muscled.

Said Bittner, "Consider this—the Patterson footage is by far the *gold standard* for Bigfoot footage. It was shot during the day, in full sunlight, out in the open on 16mm film. Independent researchers examined the location immediately after the encounter, and footprint casts and countless measurements and photos were taken. Hard to imagine better circumstances, and yet this film remains controversial, written off as an *obvious hoax* by many. So that's what you're up against. In order for a photo or video to make a significant impact, it needs to be at least as good as the Patterson film, if not better." Indeed, this film remains among the most compelling evidence for the existence of sasquatch, detractors and skeptics notwithstanding. Innovative modern techniques of image analysis have revealed new details lending further corroboration to the film.

7

THE WRANGLER AND THE WILDWOMAN:
BOB GIMLIN'S ENCOUNTER WITH BIGFOOT

Bob Gimlin is a horse rancher and builder in Yakima, Washington. The first time he had heard of sasquatch was during a trip through British Columbia, Canada, when he and his wife saw a sign for "sasquatch country" depicting a hair-covered figure. His wife suggested that it must be some kind of Indian tribe, to which Gimlin responded, "They are not very good-looking Indians." He maintains that he was agnostic on the subject of sasquatch's existence, holding a noncommittal "show-me" attitude. A friend by the name of Roger Patterson, a fellow rodeo rider, became enthusiastic over the prospect of finding Bigfoot after reading an article written by Ivan Sanderson in 1959, about America's "abominable snowman" in the mountains of northern California. Several years later, Patterson learned from tribal elders on the Yakama Indian Reservation that these huge creatures once lived throughout the Northwest and some believed they still roamed the more remote mountainous areas. In 1964, Patterson made a business excursion to Los Angeles, and stopped in northern California along the way to see what was behind Sanderson's article. He made several acquaintances, including Al Hodgson, proprietor of the general store in Willow Creek, taped interviews and made some plaster casts of footprints himself, and showed these to Gimlin. Gimlin remarked, "I was not convinced that they really existed. You know, I figured Roger must have had a reason. He showed me plaster casts and I heard different stories from people who had seen them, so I thought well maybe there is something to this, but I just didn't believe in them basically. . . didn't believe it was possible they could exist."

However, Gimlin remained open-minded. Hearing the eyewitness accounts and examining the casts, he thought there must be *something* there that he didn't understand. His interest was piqued and he accompanied Patterson to investigate reports in the Yakima region, and the two rode on horseback many times in the forests around

Mt. St. Helens. Returning from one such trip they found a message waiting for them from Al Hodgson in Willow Creek, California. Hodgson had learned from John Green about fresh tracks found along the Blue Creek Mountain road discovered after the Labor Day weekend. He waited until Green and René Dahinden had examined the tracks and left the scene and then gave Patterson a call to let him know new tracks had recently been discovered. Patterson was hopeful of getting footage of fresh tracks to include in a documentary film he had been putting together over the previous eight months, featuring interviews of eyewitnesses and shots of the locations where they had reported seeing a sasquatch. His intent was to be able to finance a full-time search for sasquatch from the proceeds of the documentary. Gimlin was between construction jobs and so he agreed to go to California with Patterson once he was able to get his affairs in order at his ranch.

Stills from what has become known as the "cowboy footage" of Bob Gimlin leading the packhorse along Bluff Creek in northern California (Courtesy of Martin and Erik Dahinden)

They took two saddle horses and a packhorse, Gimlin's one-ton truck and horse van, and supplies to keep them for two weeks or more. In fact, it was Gimlin's equipment that made the trip possible. Patterson was out of work at the time and was receiving some financial backing from his brother-in-law, Al DeAtley.

By the time they arrived, rains had effectively eradicated whatever tracks remained on Blue Creek Mountain. Although traces remained, Bob admitted he could not make much out of them. They set up a base camp down across Bluff Creek and began patrolling the dirt roads. The routine was to wait until the heavy construction and logging rigs were off the roads and then ride them on horseback, looking for tracks or whatever might turn up, returning to camp at night. By night they would slowly drive the roads in the truck looking for tracks that might have been left across them. They also rode out singly in order to cover more ground, searching for sign along the creek beds and up the mountainsides throughout the region. This went on for a couple of weeks.

Gimlin recalled the morning of October 20, 1967. "The day we got the film footage, I left early in the morning and Roger slept in. I just rode out and around, I always got up early and so I rode on out. My horse loosened a shoe and I came back in to tack the shoe on tighter. About ten, midmorning or so, I sat around there for a little while, because Roger was gone when I got back. Supposedly he had gone down the creek there—ah, Bluff Creek there—and after awhile he came back and asked what area I had covered that morning. I told him and he says, why don't we ride up into this area we had ridden into before, a desolate type area down a couple of canyons, there's a creek running through it. So we went ahead and fixed lunch and he said let's get our gear together so when we ride out we can stay if we have to and stay a little bit later into the night if we need to. We packed up the packhorse and it was about midday."

They were riding upstream on the right-hand side of Bluff Creek. Roger was in the lead, followed a horse-length or two behind by Gimlin with the packhorse in tow. Several miles upstream, they skirted a large downfall tree with a large root wad that had diverted the flow of the creek. As they rounded the obstruction, there was a logjam—a "crow's nest"—left over from the flood of '64 that had scoured the narrow valley and piled up the logs. Suddenly there was the creature standing by the edge of the creek a mere 60–80 feet to their left. In Gimlin's words, "When I first saw it, it was standing, looking straight at us. That's when everything started happening. The horses started jumping around, raising the devil and spooking from this creature. Roger, well his horse was rearing up and jumping around."

Patterson's horse, younger and less experienced, tried to spin around and come back. Gimlin's was a more seasoned roping horse but was still spooked by the encounter with the figure. Patterson was trying to control his horse with one hand while reaching back into the saddlebag for his camera with the other. He was quite agile and athletic, since he did rodeo riding and gymnastics. This was a maneuver he had practiced. "He always kept that saddlebag ready. The saddlebag had two straps on it to keep it buckled down. He kept one buckled and one of them unbuckled so he could get his camera in the event he needed it in a hurry and this was the case at that particular time . . . that was his theory that if he ever had to get it, he kept the one buckle on there so it would not bounce out while he was riding and the other one loose so he could get it out in a hurry," said Gimlin.

Patterson slid off the horse with his camera in hand and the horse ran off, prompting the packhorse to jerk free from Gimlin and follow. Patterson called out "Cover me!" as he ran across the creek toward the sandbar, which had a slight elevation of about 30 inches, the camera to his eye. With his vision restricted by the viewfinder, he ran into the sandbar and fell to his knees. Gimlin could see this within his field of vision, while keeping his eye on the creature, which had immediately turned and begun

Early frames from the Patterson-Gimlin film clip, when the subject was about sixty to eighty feet distant (Courtesy of Martin and Erik Dahinden)

retreating up the sandbar and parallel to the creek bed. Gimlin rode across the creek, dismounted, and pulled his 30.06 rifle from its scabbard. He figured if it became necessary, he could get off a surer shot on foot than in the saddle on a jittery horse. He recalled that at the time he was young, was still hunting and was an excellent shot. They always carried rifles when they rode in the mountains, but not with the intent to shoot a sasquatch. "We had talked about it, but decided unless it was necessary, we would never shoot. In other words, unless it was violent or attempted to attack us . . . I just stood there with my rifle. I never raised the rifle like I would shoot or anything like that, just held it in my hand and with the other hand held my horse to keep him from getting away from me."

The creature reacted twice to their approach, glancing back once when Gimlin crossed the creek on horseback. At that moment Gimlin was at his closest to the creature, perhaps less than sixty feet. The second glance, which has become the most publicized frame from the film (frame 352), occurred either when Patterson repositioned himself to a better vantage point and/or when Gimlin dismounted behind him. It was all happening so fast; he was uncertain precisely where the creature's attention was focused.

Fatigued from weeks of hard riding and searching the countryside day and night, the gravity of the moment did not immediately weigh upon Gimlin. Then the reality of what they were witnessing suddenly struck him. "When I saw this thing—it's almost unexplainable how I felt—I thought, is this really true? Is this really happening? Here I am, I've been here, I'm tired, but this thing is real. It's real. It's humanlike. It's walking upright and it doesn't seem to be walking fast, but it's covering the ground quickly.

"Its walk was extremely graceful, especially for a huge creature like that. I did notice that it brought its knees up fairly high, but I took into consideration that it was a pretty heavy animal—or pretty heavy creature. I'm not going to call it an animal because I don't believe it is."

Gimlin described his initial impressions of the creature's size and weight—"I thought it was about six and a half feet tall and I would have guessed its weight at 250 to 300 pounds. It did have tremendous muscle bulk. It was massive. This was an estimated guess at the time of course. I'm not used to seeing things like that. I was just guessing weight compared to the amount of muscle quarter horses have. It wasn't as big as a quarter horse naturally. And the height, because we were up on our horses at the time we first saw the creature, it probably didn't look as tall as it really was. Now the horse I was riding was a sixteen-hand horse. One hand is four inches on a horse. My horse was sixteen hands tall plus my saddle. That would make him approximately sixteen and a half hands high. Now of course, with me sitting up there, you can figure my eye level was about nine feet high. So anything actually less than nine feet you would be looking down at it."

These were Gimlin's initial impressions, but after examining the footprints deeply impressed in the firm sand compared with the shallow hoof prints left by the horses, he reconsidered. "We knew it had to be heavier than it appeared to be when we first saw it. The horse that I was riding was around 1,200 to 1,300 pounds. I rode him alongside the tracks. With new film in the camera, Roger took pictures of how deep the horse's prints were in the soil compared to the creature's tracks. Of course, we thought the horse's weight was distributed on four feet and I'm not good with the mathematics of such things, but if you figure a 1,400-pound horse distributed on four feet would be about 350 to 400 pounds, we figured it must have weighed much more than we originally figured. Then I got up on a stump, which was approximately three to four feet. We didn't measure it, probably should have. Anyway I jumped off with a high heel boot as close to the track as we could. Then we took pictures of that to illustrate the depth that my footprint went into the same dirt with a high heel cowboy boot and at that time I weighed 165 pounds . . . Course Roger did some research by going over to the zoo in Seattle; watched the gorillas there and asked how much they weighed and so forth. They had one over there named Bobo and I don't remember his weight exactly, but I do remember he weighed more than it looked like he weighed . . . In the end it probably weighed approximately 500 pounds to make tracks that deep in the dirt.

"It wasn't looking directly at me, but I could see the face real good and I could see the eyes," Gimlin recalled, describing their initial encounter with the creature. As for the details of its face—"The face would have a flat type nose; the lips, I can't really remember what the lips looked like except it did have lips and we could see its teeth. The eyes were large eyes but not big round eyes like a horse or a cow but there were large eyes. The hair on its face was short. There wasn't a whole lot of hair around its cheeks and down alongside the face. The best I can remember is the face didn't have a whole lot of hair on it."

The creature walked away with an easy motion, swinging its arms like a human. Gimlin, familiar with livestock was especially impressed by the evident and dynamic musculature. "Yes, I could see the muscles clearly and that was one of the deciding factors in my opinion that this was no 'man in a suit.' The thighs, the buttocks, the arms and shoulders, you could see it move clearly underneath the hair."

Regarding the creature's gender he continued, "Well, it appeared to be a female, but you know I had never seen one. I had never even seen a track until that day so I couldn't

Gimlin reported that the subject looked back when Gimlin rode across the creek, drew his rifle from its scabbard, and dismounted. (Courtesy of Martin and Erik Dahinden)

even make a statement whether it was male or female. But the film indicates that it had mammary glands, so we assumed it was a female."

Since the tracks found earlier on Blue Creek Mountain were of three different sizes, Patterson and Gimlin had assumed there were a male, a female, and a younger one in the area (in retrospect the three were probably a female and two offspring, perhaps even the

Roger Patterson pouring a cast at the film site and displaying the cast upon their return to Yakima, Washington (Courtesy of Martin and Erik Dahinden)

female they encountered). The prospect of possibly two more enormous creatures being in the vicinity, perhaps an even larger male, was a bit disconcerting. Gimlin was careful to ensure his horse didn't bolt from the scene. He continued, "It was moving away from me. That was about all that was in my mind at that time. That this creature was of no threat to us and oh yeah, I was trying to keep my horse under control cause you know I never had any idea what might happen and I sure didn't want to be on foot! So I knew I could get back on my horse if I had to. If I had to shoot it and it didn't go down, I could get on my horse and I could get out of there . . . and Roger would have to fend for himself. I'm not a coward, but I'll be darned if I was going to stick around if this creature got violent, you know? So I was concentrating on keeping my rifle in my hand and my horse under control."

The creature continued its course of retreat upstream disappearing around a knoll that dropped down from the canyon wall, but Patterson, still on foot and without his rifle was nonetheless uneasy about the situation. Said Gimlin, "Then when the creature did disappear up a little draw, why I wanted to follow it. Of course Roger didn't want to follow it because he was on foot and he didn't want to be left there. We thought there was the possibility there were the two others around . . . we didn't know at the time whether that was one of the ones that had made the tracks up above the scene or not. Roger was a little bit upset about that so he wanted to catch his horse and get some more film in the camera. It took quite a while to catch the horse and to catch the packhorse as well and tie them up. Then we rode on in pursuit of the creature."

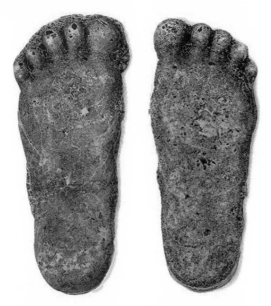

The 14.5-inch right and left footprints left in firm, moist sand, cast by Roger Patterson on the Bluff Creek film site in October 1967

Patterson quickly changed out the film under a poncho. Gimlin optimistically thought they might catch another glimpse of the creature. They followed its tracks for about a quarter mile. Beyond the sandbar they became little more than scuffs in the gravel. The stride increased to 68–72 inches, suggesting it had possibly broken into a run when it was out of sight. "Then we tracked on up the creek bed quite a ways. We saw one wet half of a footprint on a rock where it crossed the creek and it went up into the mountains and that was as far as we went with it." Gimlin was ready to continue the pursuit, even if they would have to proceed on foot up the steep mountainside. Patterson thought not. The days were short that time of year and the prospect of being out on foot in the dark dissuaded them from going farther. "He wanted to get back and take plaster casts of the tracks and then go on into town to see if we had anything on film. We weren't sure from Roger stumbling and falling down on the sandbar and getting up and running. We didn't even have an idea that we had anything on film at that time. In fact, it was doubtful that we did have anything." That was the object in getting the film back to DeAtley as quickly as possible for development and then get-

Still image of a footprint on the Bluff Creek sandbar from the second roll of film (Courtesy of Martin and Erik Dahinden)

ting back to the site with more film and calling in the tracking dogs to pick up the trail where it went up the mountain, to see if more footage could be obtained. Gimlin recalled Patterson lamenting, "I don't know, Bob . . . I'm not sure that I got any film footage at all—I fell down, got relocated, then got relocated again, and then I ran out of film!"

They returned to the truck back at camp and retrieved some plaster. They poured casts of two clear footprints—a right and a left. Casting and filming the tracks occupied what remained of the afternoon. It was nearly dark by the time they arrived at camp and got the horses tied and fed. They got in the truck and drove through Weitchpec and over the Bald Hills road to the coast at Orick, then down to Eureka where Roger shipped the film to DeAtley for developing.

They returned through Willow Creek and stopped to see Al Hodgson, proprietor of the local general store, who held a personal interest in the topic, phoning him after hours from in front of the store. Hodgson said he took the call and heard Patterson say, "I got a picture of the son-of-a-buck!" Hodgson had taken Patterson as always a calm speaking sort of man, but on this evening Patterson was quite excited. Hodgson observed, "Honestly, he was three feet off the ground. He was very, very excited. Afterward I turned to my wife, Frances, and asked her what she thought. She said either he saw it or he's on LSD." Hodgson called Syl McCoy from the Forest Service and they talked for several hours. Afterward Hodgson recalled, "I'm not an expert in interrogation, but as far as we could tell, he [Patterson] was telling the truth." Patterson and Gimlin went to buy some groceries and before leaving, Patterson asked Hodgson to contact Green and Dahinden, two Canadian investigators from British Columbia, about arranging for tracking dogs.

They arrived back in camp late that night and sat up for several more hours discussing what they had seen, especially what Gimlin had noticed, as he had the better view, while Patterson was looking with one eye through the camera's viewfinder. Sleep that night was brief, interrupted when Gimlin awoke to the pitter-patter of rain, which quickly increased in intensity. "Around 5:30 A.M. or so it started raining, and it was just a pouring-down rain. I told Roger we better get up and do something about the tracks or they'd wash out, and he said no, it would stop raining after a while. I went ahead and got up, put the saddle on my horse, and decided I would ride up there while it was raining really hard. I had gotten a couple of cardboard boxes from Mr. Hodgson's to cover these tracks the night before. So when I went outside to get a couple of these boxes that were folded up out there, they were just soggy old pieces of cardboard. I disregarded taking those back up there. So I rode back up to the scene, pulled some bark off some dead trees, and covered up the tracks as best I could, the best ones that I could identify. In the rain like that it had already done some damage to them. Then I went back to camp. By then we decided it wasn't going to quit raining. The little creek that was six or seven feet across was now ten or twelve feet across and four feet deep! We were on the side of the creek, which had to be crossed with the truck to get out to the main road. I said, 'Well I'm going to go ahead and cross the creek with the truck and get started out.' And of course Roger thought it would stop raining and he suggested I leave him there and come back and pick him up." Gimlin's truck was a two-wheel-drive one-ton and he felt they needed to get moving. The rain virtually eliminated any prospects of using the tracking dogs, even if they could get to the site and pick up the trail into the mountains. Now Gimlin's concern was getting his rig out of the mountains. The rain didn't let up and the low road was blocked by a flowing mudslide several feet deep. Trees were sliding down off Onion Mountain. They nearly backed their vehicle into the creek trying to turn around. The windshield wipers couldn't keep ahead of the torrent and visibility was poor. It took nearly all day to get back to the main highway. They drove all night to get back to Yakima.

Eventually, Gimlin relived the experience by viewing the developed film footage. Of this he said, "I was kinda disappointed with the original film, because I thought it should have gotten a lot more than that, but with Roger running part of the time and trying to relocate part of the time and these things happened so rapidly—it was just happening in heartbeats . . . I was relieved to see that he got that much. At least there was something there; that you could see the movement and you could see the muscles." Gimlin added a remark that also sheds light on the question of the film frame speed controversy that arose later. Patterson, who usually kept the frame speed set at 24 frames per second (fps), discovered the snap dial positioned at a slower setting, 16 fps. This had fundamentally significant implications for the estimation of the height of the subject and the characterization of its walk, but Patterson confessed that he simply had no idea when the dial had been reset. Quite pertinent to this matter is Gimlin's observation, "What didn't please me was that I'd see other people run the film footage [on TV] maybe at different speeds or what [typically 24 fps], but it didn't look the same, it didn't even walk the same. I was terribly disappointed with that."

Some twenty-five years later, Gimlin reflected, "My impression is that there is a creature, and I don't feel it was a man in a suit. If it had been a man in a suit, I don't know how they would have gotten him back into that particular area. I have heard this story and thought about it many times. God! At one point with the film circulating all around and people criticizing, I was almost to the point of not being even sure myself. But I thought about it all these years and I'm quite sure it wasn't a man in a suit. I saw the face. I saw the expression on its face. With all the muscles in the arms and legs, I don't know how it could be a man in a suit! Plus I never had anything to do with a man in a suit and *if* Roger did, how would he know I wouldn't shoot it? In my opinion, that creature was not a man in a suit." Although the two had previously agreed not to shoot a Bigfoot unless their lives were at risk, who's to say precisely what circumstances might constitute a risk in any one person's estimation placed in such an unpredictable situation. Gimlin couldn't imagine Patterson putting him in that position.

At the Willow Creek Bigfoot Symposium, held in 2003, Gimlin was asked if he ever regrets his involvement in the incident. He replied that he had no regrets then and he

On the occasion of the Willow Creek Conference held in 2003, Bob Gimlin returned to the film site for the first time. Pictured from left to right are: author, Bob Gimlin, John Bindernagel, Daniel Perez, John Green, and Dmitri Bayanov. (Courtesy of Daniel Perez)

had none now after meeting so many nice people at the conference. There had been times, however, when he did wish he had never been involved. He could handle the criticism, but his wife had also endured a lot of ridicule. At times it was a strain on their relationship. In his words, she is not as thick-skinned as he, and so was more sensitive about the ridicule and harassment they experienced. In spite of the incredulous reaction to the film by many, Bob Gimlin reiterates that there is no doubt at all in his mind that he had seen a real animal. Once again he calmly and confidently reiterated, "They are real; I saw one."

How good is the Patterson-Gimlin film? If authentic, this is one of the most startling natural history films in existence. It has received the most extensive forensic analysis by anthropologists, zoologists, and film experts of any of the purported films or videos of sasquatch. What do the analyses of the experts say about the dimensions, anatomy, and gait of the creature captured in this footage?

8

Picture This: Scientific Reaction to the Patterson-Gimlin Film

Patterson was eager to rush off to show the film to experts in Los Angeles and New York; however, Green and others argued that he would lose credibility away from his own setting. At length he was persuaded to first show the film at the University of British Columbia, Canada, where a few of the resident authorities had already been exposed to some of the evidence of sasquatch. As might be expected, reactions by the scientists were at best guarded and essentially noncommittal.

Dr. Ian McTaggert-Cowan, a leading zoologist, summarized the ingrained bias directed at the question of sasquatch, saying, "The more a thing deviates from the known, the better the proof of its existence must be." This declaration has often been rehearsed by those who are prompt to dismiss *any* evidence of sasquatch. Justified or not, this approach seems to apply a double standard to evidence of sasquatch. It would seem that an equally rigorous standard of evaluation should be applied to all evidence laid out for scientific consideration, regardless of preconceptions, since history has attested that perceptions based on yesterday's understanding are often of necessity revised in the light of today's revelations.

Don Abbott, curator of anthropology at the Royal British Columbia Museum, who had just weeks before examined firsthand the extensive trackways of three individual sasquatch in the region of Bluff Creek represented the feeling of a number of scientists who were duly impressed by the film but found it a challenge to abandon their preconceptions. He said, "It is about as hard to believe that the film is faked as it is to admit that such a creature really lives. If there's a chance to follow up scientifically, my curiosity is built to the point where I'd want to go along with it. Like most scientists however, I am not ready to put my reputation on the line until something concrete shows up—something like bones or a skull."

Don Abbott examining footprints on the Blue Creek Mountain Road in northern California (Courtesy of John Green)

Dr. G.C. Carl, the provincial museum director, was sufficiently impressed to make a promise to scientifically investigate any fresh tracks that turned up in British Columbia.

Dr. W.J. Houck, of Humboldt State College in Arcata, California, attended the viewing in B.C. as a complete skeptic, but left undecided. "I'm not going to call it a hoax," he told the Province newspaper, "yet the alternative is still too fantastic to accept. Where does that leave me? Darned if I know."

After a day or two of notoriety in the regional press it became apparent that nothing further could be accomplished locally. Inertia had prevailed. However, *Life* magazine expressed interest in the story and therefore arranged for another showing to scientists, this time at the American Museum of Natural History in New York City. In attendance were Dr. Richard Van Gelder, head of the Department of Mammalogy and Dr. Harry Shapiro, from the Department of Anthropology, along with members of the press. During a brief showing of the film, Patterson and DeAtley were left waiting outside the conference room. The expert consensus was, "It is not kosher because it is impossible." In other words, it was considered impossible that a giant primate could exist in North America yet unknown to science; therefore, the only explanation for the film was that it was a hoax. As a result of this negative assessment *Life* washed its hands of the matter and suddenly dropped out of negotiations.

Argosy magazine promptly took on the story and arranged a second U.S. showing, this time at the Smithsonian, in Washington, D.C., which was attended by Dr. John Napier, who was director of the Primate Biology Program at the time, Dr. Joseph Wraight, Chief Geographer, U.S. Coast Guard and Geodetic Survey (a human ecologist), Dr. Vladimir Markotic, associate professor of archeology at the University of Calgary (a physical anthropologist), and Dr. Alan Bryan, professor of anthropology at the University of Alberta. Sanderson wrote the article for *Argosy* and quoted favorable comments by Napier and Wraight. Markotic and Bryan were not quoted, but Markotic considered the film genuine and was involved in further investigations until his death.

It was not only a men's magazine that publicized the story. The *National Wildlife* magazine, the outlet for the National Wildlife Federation, promptly picked up the story as well. An article penned by executive editor and experienced outdoorsman Dick Kirkpatrick, presented a balanced matter-of-fact treatment of the story to wildlife enthusiasts and conservationists. He concluded with his reflections on just what the film depicted. "I don't know. After several weeks of research and interviews, I'm as confused as I was in the beginning," he said. "It could be a big, hairy anthropoid creature, rare enough and shy enough that it simply hasn't been captured yet. Where there is that much smoke, there just may be some real fire." Shortly thereafter, the January 1969 issue of *Reader's Digest* carried a story by James B. Shuman, "Is There an American Abominable Snowman?", condensed from *West* magazine, a publication of the *Los Angeles Times*, which reached an even wider readership.

Some years later, Dr. Richard Thorington, Jr., who succeeded Napier as director of the Primate Biology Program at the Smithsonian, commented on *his* rather contrasting assessment of the reaction to that historic viewing of the Patterson-Gimlin film: ". . . To them it appeared all too obvious that the pictures were made of a person dressed up in an ape costume, trying to run in an unnatural way. The one person who did not at the time consider this to be the case was Dr. John Napier . . . let me point out that there are two ways to look at the 'Bigfoot phenomenon.' One is the approach that Dr. Napier espoused that we should keep an open mind and review all evidence to decide whether this is a hoax or a legitimate area of study." Dr. Napier is thus incorrectly singled out as the only expert at the viewing who lent any credence to the film, let alone a willingness to give the evidence a reasonable hearing. Thorington went on to suggest that the only tenable position—one that he felt "forced to espouse"—is ". . . that one should demand a clear demonstration that there is such a thing as Bigfoot *before* spending any time on the subject. There are many, many valid areas of research for which the subject matter is known to exist, so one should busy oneself with these rather than with will-o'-the-wisps." If such an attitude prevailed without exception throughout science, it would be a wonder that any breakthrough discoveries were ever realized.

Dr. William Montagna, former director of the Oregon Regional Primate Research Center, was among those viewing the film and reflected on his experience in *Primate News,* September 1976: "Along with some colleagues, I had the dubious distinction of being among the first to view this few-second-long bit of foolishness. As I sat watching the hazy outlines of a big, black, hairy man-ape taking long, deliberate human strides, I blushed for those scientists who spent unconscionable amounts of time analyzing the dynamics, and angulation of the gait and the shape of the animal, only to conclude (cautiously, mind you) that they could not decide what it was. For weal or woe, I am neither modest about my scientific adroitness nor cautious about my convictions. Stated simply, Patterson and friends perpetrated a hoax. As the gait, erect body and swing of the arms attest, their Sasquatch was a large man in a poorly made monkey suit. Even a schoolchild would not be taken in. The crowning irony was Patterson's touch of glamour: making his monster into a female with large pendulous breasts. If Patterson had done his homework, he would have known that regardless of how hirsute an animal is its mammary glands are always covered with such short hairs as to appear naked.

"To believers who claim that we scientists are too opinionated to look at the evidence, I reply: Is a scientist to listen to every zealot who regales him with tales of a putrid stench, who shows him fake footprints, or makes films of a man wearing a badly tailored monkey suit? The scientist who is reviled because he won't listen to fantasy goes securely on his way, knowing that life is so full of real wonderment and mystery that he does not have to fantasize. But perhaps I ought to add that man's need to fantasize is a vestigial remnant of his past. It created mythological characters, good and evil; visions and miraculous events, heaven, purgatory and hell. It created the oracles, the art of palmistry, phrenology, astrology, and all sorts of other occult sciences. And finally it peopled man's world with monsters."

The smug confidence in Montagna's assessment of the situation, even some years later, lends little to his credit. His often-repeated statement about the contradiction of hair-covered breasts is one that has always puzzled me. First, the degree of hairiness of the breasts of the film subject is open to interpretation because the exposed areas of skin are nearly the same color as the hair, excepting the appearance of the soles of the feet, and the palms of the hands, a matter addressed a bit later on. The resolution of the film makes it difficult to be certain how densely covered with hair are the presumed breasts. On the other hand, anyone has but to peruse the many published photographs of apes to discover that the ape's chest *is* covered in hair, no matter how fine. In a study by Wettstein, a sample taken from the middle of the chest of a series of apes yielded the following average hair densities per square centimeter: Gibbon 600, Orangutan 100, Chimpanzee 70, Gorilla 5, and *Homo* 1. Human female breasts are indeed covered with hair follicles and associated erector pilli muscles that cause goose bumps, although the hair is extremely short and fine. This hair is sensitive to masculinizing androgens, circulating sex hormones that cause the hair to become heavier and darker in typical areas, including the chest. With this in mind, and given the evident variation among the closely related hominoids, where is Montagna's justification in drawing such a dogmatic conclusion about the appropriateness or otherwise of the apparent hairiness of sasquatch breasts?

One of the first persons to take the initiative to investigate the alleged film site and the apparent circumstances of the encounter was Bob Titmus. In a letter to John Green, Titmus explained his reaction and his impressions of the incident:

> By the end of the day it had become apparent that a few of the viewers felt there was the possibility the whole thing was a very elaborate and expensive hoax. I felt that this possibility was so extremely remote as to be almost nonexistent. (None of these individuals witnessed more than one showing, I believe.) However, I did have to take into consideration the fact that I believe that I viewed the film through somewhat different eyes than most of the persons present.
>
> First, I think that a taxidermist will see and retain far more detail, while watching an animal, and is probably far more qualified to recognize anything unnatural, than the average person.
>
> Secondly, evidence that I witnessed in the mountains of Northern California about ten years ago changed me from a nonbeliever to a believer and since that time I have spent a major portion of those years, as you know, interviewing witnesses, investigating reports, collecting evidence, casting many, many different tracks, setting up camera and live traps, tracking the creatures dozens of times, etc., all of this was in an effort to capture one of the creatures. All of this only strengthened the case of the existence of the creature Bigfoot/Sasquatch.
>
> Thirdly, many years ago I saw one of these creatures at fairly close range and watched it for about ninety seconds before it walked off into the timber.

Since I know more about tracks than film and generally feel that they will tell me a more accurate story than film, I had a very strong urge to see the tracks that were being made during the time that Roger was shooting his film. I felt that the tracks could very well prove or disprove the authenticity of the pictures. No one else present seemed inclined or able so the following day I went on to California to have a look at the tracks.

My first full day up near the end of Bluff Creek, I missed the tracks completely. I walked some fourteen to sixteen miles on Bluff Creek and the many feeder creeks coming into it and found nothing of any particular interest other than the fact that Roger and Bob's horse tracks were everywhere I went. I found the place where the picture had been taken and the tracks of Bigfoot the following morning. The tracks traversed a little more than three hundred feet of a rather high sand, silt, and gravel bar, which had a light scattering of trees growing on it, no underbrush whatsoever, but a considerable amount of drift debris here and there. The tracks then crossed Bluff Creek and an old logging road and continued up a steep mountainside.

This is heavily timbered with some underbrush and a deep carpet of ferns. About eighty or ninety feet above the creek and logging road there was very plain evidence where Bigfoot had sat down for some time among the ferns. He was apparently watching the two men below and across the creek from him. The distance would have been approximately 125–150 yards. His position was shadowed and well screened from observation from below. His tracks continued on up the mountain, but I did not follow them far. I also spent little time in trying to backtrack Bigfoot from where his tracks appeared on the sandbar since it was soon obvious that he did not come up the creek but most probably came down the mountain, up the hard road a ways and then crossed the creek onto the sandbar. It was not difficult to find the exact spot where Roger was standing when he was taking his pictures and he was in an excellent position.

I spent hours that day examining the tracks, which, for the most part, were still in very good condition considering that they were nine or ten days old. Roger and Bob had covered a few of them with slabs of bark etc., and these were in excellent condition. The tracks appeared perfectly natural and normal. The same as the many others that we have tracked and become so familiar with over the years, but of a slightly different size. Most of the tracks showed a great deal of foot movement, some showed a little, and a few indicated almost no movement whatever. I took plaster casts of ten consecutive imprints and the casts show a vast difference in each imprint, such as toe placement, toe gripping force, pressure ridges and breaks, weight shifts, weight distribution, depth, etc. Nothing whatever here indicated that these tracks could have been faked in some manner. In fact, all of the evidence pointed in the opposite direction.

Bob Titmus seated in front of his cast collection, including ten successive footprint casts made at the Patterson-Gimlin film site, which are now held at the Willow Creek–China Flats Museum (Courtesy of John Green)

And no amount of thinking and imaging on my part could conceive of a method by which these tracks could have been made fictitiously.

His sister Allene and brother-in-law Harry Halbritter joined Titmus at the site. "Harry has hunted big game all of his life. He has been all over Africa, Alaska, Yukon Territory, Canada, Mexico, and the United States and stated that this impressed him more than anything he had ever seen in the bush in all of his travels. Harry made several tests and observations, one of which was to walk briskly beside the tracks to try to match their depth of up to an inch and a quarter in places. Harry is a two hundred-pounder and the best he could do was an imprint of about a quarter of an inch on the rear portion of his shoe heel, and one eighth of an inch and less on the rest of his shoe imprint. We both agreed considering the depth of the two imprints and the difference in the amount of the bearing surface that the creature that made these tracks would have to weigh at least six hundred to seven hundred pounds."

The next summer John Green visited the site and conducted a simple and straightforward size comparison of the Patterson-Gimlin film subject. The remains of the trackway were still evident as a series of rough depressions stretching across the sandbar. Equipped with slides from the movie he was able to relocate Patterson's and the film subject's approximate positions, by lining up objects through the viewfinder of a similar camera. Although inexact, this method could at least offer an indication of the estimated size of the subject. Green filmed 6' 5.5" Jim McClarin, a student at nearby Humboldt State University, as he retraced the path followed by the alleged sasquatch.

John Green filmed 6' 5.5" Jim McClarin retracing the path walked by the film subject. Precise superimposition of the frames yields this revealing split-screen image. (Courtesy of John Green and Martin and Erik Dahinden)

156   SASQUATCH

The footprints were still evident as vague depressions, the trackway punctuated at points where the effects of time had erased them altogether. McClarin had been on the scene with Richard Henry shortly after the original events on November 5, 1967, when the trackway was still very clear and he had become quite familiar with the specific course taken by the subject across the sandbar. In an interview with Daniel Perez, an amateur investigator (and author of the definitive publication about the provenience of the film), Henry noted an interesting detail of the dynamic nature of the tracks. He described the high bank on the far side of the creek, marking the edge of the elevated sandbar traversed by the film subject. The bank was 28–30 inches high. The film subject took the bank in one step, cresting the bank with only the ball of the foot and the toes, which slid down digging into the edge of the bar. Both were impresssed with the natural appearance of the trackway.

When the corresponding frames of Green's film of McClarin were placed over the Patterson-Gimlin film, the alignment of features in the foreground and background was quite correct, as well as the resulting superimposition of the steps of the two individuals filmed. If there was any apparent discrepancy, it was that John held the camera slightly lower on occasion. In those instances, the camera angle would make McClarin look slightly taller in relation to the subject. But at the distances involved, this discrepancy would be negligible. From this demonstration it was evident that the creature stood taller than the 6' 5.5" McClarin. Even more distinctly evidenced in the comparison was the subject's massive build, which dwarfed McClarin, who at the time weighed 180 pounds. The breadth and depth of its torso, shoulders, and arm span were considerably greater than McClarin's.

In 1969 John Green met with Ken Peterson, a senior executive at Disney Studios, who told him that their technicians had examined and studied the film. They concluded that in order to create something like it they would have to resort to animation. None of their animitronics was sophisticated enough to move freely. According to them, if it was hoaxed, it had to be a man in a suit. One of the top men in the costume field at that time was Janos Prohaska, a Hungarian-born stuntman and costume designer in Hollywood since 1939. Prohaska's ape characters made costumed appearances during the late 1960s in such television series as *Gilligan's Island, Perry Mason, Land of the Giants, Here's Lucy,* and *Red Skelton,* to name a few. One of his most memorable appearances was as the white unicorned gorillalike *mugato* that bit Captain Kirk on the *Star Trek* episode, "A Private Little War" (1968). After studying the film, Prohaska was convinced that it depicted a living creature rather than a man in a suit. He noted, "You could see all the muscles on the body . . . It didn't move like a costume at all." He felt that instead of a suit it would have to have been a minimum ten-hour makeup job in which the hair was glued directly to the actor's skin. He concluded, "It looked to me very *very* real. If that was a costume that was *the* best I have ever seen." Isn't it curious that such a hypothetically skilled costume designer had never been employed in the Hollywood film industry then or since?

No American or Canadian scientists were inclined to undertake a systematic analysis of the film, so René Dahinden, a long-time investigator, took the film across the Atlantic to have it studied. He succeeded in recruiting the expertise of Dr. Don Grieve, a professor of biomechanics at London's Royal Free Hospital of Medicine, and Dr. Dmitri Donskoy, chief of the chair of biomechanics at the USSR Central Institute of Physical Culture in Moscow. The complete reports of each of these analysts have been published elsewhere, but they deserve further comments here. A number of Grieve's metrics must be qualified due to a handicap. He was not provided with the length of the footprints or the measured step. These he derived from the film itself and came to the conclusion that the foot was 13.3 inches long, instead of nearly 15 inches. In spite of this he summed up his overall impressions by saying, "My subjective impressions have oscillated between total acceptance of the sasquatch on the grounds that it would be difficult to fake, to one of irrational rejection based on the emotional response to the possibility that the sasquatch actually exists. This seems worth stating because others have reacted similarly to the film."

The ultimate conclusions of Grieve's analysis hinged upon the frame speed. He observed that if it was shot at 24 frames per second (fps) the quantitative details of the gait were reasonably similar to a human. However, if the film were shot at 16 or 18 fps the possibility of a human mimicking the distinctions of gait were virtually eliminated. Grieve said, "The possibility of fakery is ruled out if the speed of the film was 16 or 18 fps. In these conditions a normal human being could not duplicate the observed pattern, which would suggest that the sasquatch must possess a very different locomotor system to that of man."

Patterson stated that he normally set the camera at 24 fps, but that after the film had been taken he found the camera set at 18 fps. He could not say when the setting might have been altered. Up to his death he maintained that he simply did not know what frame speed the camera had been set at during the filming. The camera's lowest settings are only marked at 16 and 24 fps. There actually is no marked setting for 18 fps, and it was assumed that Patterson simply misspoke or was unfamiliar with the rented camera's settings when he stated that he found the dial set to 18 fps. The confusion led to analyses being conducted at all three frame speeds. Dr. Grover Krantz's analysis indicated that 18 fps was the probable speed although the camera dial is a snap dial, precluding intermediate settings between 16 and 24 fps. However,

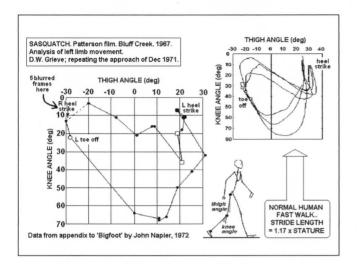

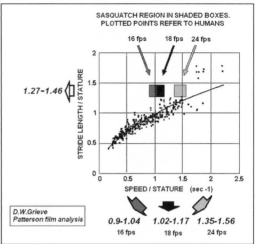

Calculations by Don Grieve indicate it was very unlikely that the gait pattern of the film subject was executed by a human. (Courtesy of Don Grieve)

the instructions for this camera model (Cine-Kodak K-100 16 mm, hand wound) note that the actual speed of individual cameras will vary within 10 percent of the indicated speed setting. Krantz related that a BBC cameraman obtained an example of the camera model Patterson used and timed the frame rate electronically when set at 16 fps and discovered that it was actually filming at 19 fps, nearly a 19 percent deviation. Therefore the potential accuracy of Krantz's conclusion of 18 fps, and its implications, are quite justified.

Several lines of evidence conclusively point to the slower frame speed. First, and perhaps most fundamentally, the degree of motion blurring evident in the film footage points to a slower frame speed. At 24 fps motion blur would be all but eliminated. Second, one of the Russian observers, Igor Bourtsev, noted that the vertical vacillations of the film frame most likely corresponded to the running or walking of the cameraman, presumably Patterson. Bourtsev observed intervals of 4 and 7 frames. He also noted that world-class sprinters manage 5 steps per second. If the film speed were 24 fps, then Patterson would have to had managed 6 steps per second to produce the vertical oscillations at 4-frame intervals, which is simply impossible. At 18 or 16 fps he was taking 4.5 or 4 steps per second, either of which are possible considering Patterson's short stature.

Krantz had employed an elegantly straightforward approach to determining the frame speed. He noted, as had Grieve, that the leg-to-stature ratio of the film subject was comparable to an average human. Since the limbs swing essentially like modified pendulums, then the period of the swing is proportional to the square root of their lengths. Therefore, the frame speed, not to mention the height of the film subject, can be determined by comparing its rate of step with a human of known height. Krantz, at 6' 5" tall, used himself as the comparative biped and found that a two-step stride required an average of 1.15 seconds if he was walking very briskly, and 1.3 seconds if he walked more slowly. If the frame speed was 24 fps, the subject's walking speed was less than one second per stride, which would be a very fast unnatural movement for someone of comparable height and would produce a jerky halting gait. At 18 fps the film subject was moving at 1.25 frames per stride, intermediate between Krantz's walking speeds, indicating that the slower film speed was the correct one. Finally, Patterson had shot film footage just prior to the encounter, showing one of them riding and leading the packhorse up the creek bottom. When played back at 24 fps the gait of the horse is very jerky and unnatural, while at 18 fps the movements appear quite appropriate, further indicating that the camera was set at the slower frame speed.

I recently presented this new information about frame speed to Dr. Grieve and asked what implications this might have for his conclusions about the film. He reiterated his position that the possibility of fakery was *ruled out* at the slower frame speed. He qualified that statement by acknowledging that it might be conceivable for a practiced actor or a dancer to execute such an extravagantly atypical pattern of gait at 16 or 18 fps, but took it as highly improbable.

Dr. Dmitri Donskoy worked out training methods for Soviet athletes preparing for international competition, including the Olympics. His assessment of the film subject is a qualitative appraisal of the synergy of the whole organism. He concluded by noting an expressiveness of movement, indicating a motor system well adapted to the task it is called upon to perform. "In other words, the neat perfection is typical of those movements which through regular use have become habitual and automatic . . . At the same time, with all the diversity of human gaits, such a walk as demonstrated by the creature in the film is absolutely nontypical of man." This is admittedly a subjective assessment that may be trivialized by some, but one that is based upon an intimate and extensive familiarity with human athletic movements and the underlying functional anatomy.

Primatologist John Napier examines a cast of a sasquatch footprint. (Courtesy of Martin and Erik Dahinden)

In 1972, Dr. Napier, one of the few U.S. scientists of very applicable expertise who maintained an open mind toward the subject, published a book titled, *Bigfoot: the*

Yeti and Sasquatch in Myth and Reality, containing his evaluation of the Patterson film, including his initial impressions upon viewing it at the Smithsonian in 1967 and after dozens of viewings both frame-by-frame and from printed stills. His assessment is remarkably at odds with those by others who have studied the film, such as the preceding statements by Dr. Donskoy. Napier concludes, "There is little doubt that the scientific evidence taken collectively points to a hoax of some kind. The creature shown in the film does not stand up well to functional analysis. There are too many inconsistencies; yet no scientist to whom I have spoken and who has seen the film has any direct evidence to prove that the episode was anything other than what it purported to be. My own comment quoted in an article in *Argosy* magazine (February 1968) was that there was nothing in the film, which would prove conclusively that this was a hoax. In effect, what I meant was that I could not see the zipper; and I still can't. There I think we must leave the matter."

Rather than leave it there, a closer examination of Napier's observations indicates that his conclusion has little foundation. He cited several impressions made at the 1967 showing at Washington, D.C., that remained unchanged by further examination of the film. His impression was that the gait was consistent with a human, specifically a male human. However, a relatively small-brained ape would not have the obstetric constraints of a human female pelvis, and therefore would not be expected to exhibit a dramatic gender difference in style of gait. He added that the walk was "self-conscious" which he went on to attribute to a conscious effort to exaggerate the stride on false feet. This inference appears at odds with his earlier observation that, "The physical and psychological attitudes of the creature appear totally relaxed almost to the degree of nonchalance." He speculated that the center of gravity appeared to be positioned comparably to a human male's, although the upper body bulk seemed to dictate a higher position of the center of gravity. Since the film subject's upper body bulk appears proportionately matched by that of the lower body, I can see no inconsistency on this point. Also, there was Napier's discomfort with the combination of apelike and humanlike attributes. Again, our present knowledge of the course of hominid evolution renders this concern obsolete; early hominids display a mosaic of apelike and humanlike traits. Finally, Napier suggested that the feet were too large for the estimated stature and the reported step length too short. This conclusion was arrived at using standardized human coefficients (without allowance for expected deviations) applied to a purportedly unknown hominoid ape, surely the application of a standard of debatable relevance. As will be discussed in chapter 13, the proportions of the foot segments are not typical of human feet, but yet are quite reasonable given the inferred mechanics of the sasquatch foot and size of the subject.

Composite illustration of several frames of the Patterson-Gimlin film superimposed with a six-foot-tall subject (Courtesy of Martin and Erik Dahinden and NASI)

After unsuccessful attempts to recruit disinterested technical analysis of the film, Dr. Grover Krantz undertook an analysis of his own, published in his book titled, *Bigfoot Sasquatch Evidence*. Not only did he have access to additional information regarding the provenience of the film, but he also came to personally know both Roger Patterson and Bob Gimlin very well. The technical aspects of the film subject, combined with the circumstantial evidence of the encounter, led him to conclude with a high degree of certitude that the film was genuine.

In another study conducted at the film site, Peter Byrne photographed a model holding a graduated staff exactly where the film was shot. The model, Michael Hodgson, was 6' and 150 pounds. Jeff Glickman, a certified forensic examiner, superimposed the photograph over frames from the Patterson-Gimlin film and aligned visible landmarks. He concluded that the film subject was 7' 3.5" tall (± 1 inch). Glickman summarized his findings in a report prepared by the North American Science Institute (NASI). He went on to note a number of anatomical details that appeared similar to those of a mountain gorilla. The congruence of the appearance of the film subject's foot and the footprints cast at the film site was demonstrated. This was noteworthy since some had contested that the footprints were hoaxed separately, after the filming. He also drew attention to an interesting detail of the visible presence and action of the kneecap. The range of motion of the knee joint was demonstrated to be distinct from typical human gait patterns. Glickman pointed out clear surface plasticity in the side of the torso as the subject turns to look back at the witnesses. The digitized film was subjected to an edge detection algorithm but no seams or borders were found. In conclusion he stated,

"Despite three years of rigorous examination by the author [Glickman], the Patterson-Gimlin film cannot be demonstrated to be a forgery at this time."

Dr. David Daegling, a physical anthropologist at the University of Florida specializing in the primate chewing apparatus, and Dr. Daniel Schmitt at Duke University, a specialist in primate locomotion, undertook a recent analysis of the film, addressing two fundamental issues: Are the film subject's body dimensions outside the range of human variation? Can a human duplicate the gait of the film subject? The results of their investigation were published in the *Skeptical Inquirer* (May/June 1999) and reiterated in a book by Daegling, *Bigfoot Exposed*. After pointing out the potential sources for measurement error plaguing the efforts to estimate the precise dimensions in the film, they flatly state that the *exact* dimensions are unknowable. On this point I must concur. However, they acknowledge that the scale of the footprints, which are correlated with the sole of the foot quite visible in the film, is known. This potential scalar is one-fifth to one-sixth of the subject's height and while not ideal, provides a reasonable reference for approximate height estimation, even as we acknowledge the potential for measurement error. The inherent scale of the foot was the primary basis for one of Dr. Krantz's estimates of stature and dimensions. Daegling and Schmitt report, "Krantz's estimate of the film subject's stature is 6' 6" (198 cm), well within human limits, but he argues that the chest width of the subject is incompatible with the human form: 'I can confidently state that no man of that stature is built that broadly.' Assuming that these parameters are measured without error, this assertion may be refuted by a quick consultation of the *Anthropometric Source Book* (1978). Chest width is measured by Krantz in the same fashion as a distance known as *interscye* in the anthropometric literature. In a sample of 1,004 men of the German Air Force, interscye of the ninety-fifth percentile is 49.6 cm, a good 3 cm larger than Bigfoot's impossibly wide thorax. The ninety-fifth percentile stature is 187.1 cm in this group, less than 4 inches shorter than the film subject. Unless Krantz would argue that taller Air Force personnel necessarily have narrower chests, his confident statement is admirable for its faith but not its veracity."

It seems their consultation of this reference was too *"quick,"* as Daegling and Schmitt have unfortunately misstated their case. By referring to the *Anthropometric Source Book,* one discovers that the "interscye" measure (#506) is "a *taped* measure across the back between the posterior axillary folds at the lower level of the armpit." A taped measure is distinguished as an arc or a circumference measured on the surface of the body, rather than a straight projected length, as would be estimated from a film image. In this case interscye is an arc following the contour of the back from one armpit to the other. If the radius of curvature of this arc is large, then the difference between the length of the arc and its cord length, or linear distance between its end points, is minor, and perhaps justifiably negligible. If there is considerable depth to the torso, the radius of curvature will be considerably smaller and the disparity between arc length and cord length can be significant. The correct measure found in the *Anthropometric Source Book* for comparison to Krantz's chest breadth is, not surprisingly, "chest breadth" (metric #223), defined as "the breadth of the torso measured at the nipple line" (without pressure). This is a straight-line measure of the greatest breadth of the torso taken at the level of the male nipple. Whether projected from the ventral or dorsal surface is irrelevant. In this dimension the ninety-fifth percentile of the same sample of German aviators measures a mere 35.4 cm, much less than the 49.6 cm value for the interscye, and also much less than the 46.5 cm chest breadth estimated by Krantz for the film subject. Indeed, the ninety-ninth percentile of the German sample still only reaches 37.4 cm, still over 9 cm shy of the film subject's estimated chest breadth. The largest reported value actually comes from a sizeable sample of 2,984 U.S. law enforcement officers, with a ninety-ninth percentile value of 41.7 cm for chest breadth and 192.6 cm for stature.

Broad-shouldered men often have a decidedly V-shaped torso; that is, their hips are not nearly as wide as their chest and shoulders. The torso of the film subject appears as wide through the hips as it does through the chest, roughly 50 cm according to Krantz's estimate of stature. Turning again to the *Anthropometric Source Book* we find that for standing hip breadth, the sample of German aviators referred to previously yields a representative ninety-fifth percentile value of 38.3 cm, once again considerably less robust than the film subject. Krantz estimated the film subject's shoulder breadth at 71.6 cm, compared to the German aviators ninety-ninth percentile value of 51.8 cm. So much for the claim of refuting Bigfoot's "impossible" dimensions. It appears that the plausibility of Dr. Krantz's estimates remain intact.

Turning to the second issue, Daegling and Schmitt take exception to Krantz's opinion that based on the manner of walking there is "no possibility" that the subject is a man in a fur suit. They acknowledge that the apparent gait is atypical of normal human walking, but point out that humans can adopt a compliant gait that approximates the deeper flexion of knees and hips exhibited by the film subject and that it also results in a longer stride and faster walk. This simplified generalization is correct, and a refined examination will certainly shed light on the distinctions between a typical human gait, a human compliant gait, and the compliant atypical gait exhibited by the film subject. However, this generalization also glosses over several aspects of lower limb movement that appear distinctive about the film subject's walk, such as the noticeably abducted foot during the swing phase. Furthermore, they note that statements by Dahinden and Perez, "that, on one of his numerous visits to the film site, he and others present were incapable of walking the distance traveled by the film subject in the time that it did so (assuming a film speed of 24 fps)" seem to attribute superhuman capabilities to the film subject. Actually the observation was intended to point out what had been determined independently—that the film could only have been shot at a slower frame speed. They did not intend to suggest that the subject was walking at extraordinary speeds. Indeed, Daegling and Schmitt present data indicating that human subjects of appropriate stature, when using a forced high-speed compliant gait, can match the film subject's walking speed, assuming 24 fps. However, this assertion trivializes the naturalness of gait exhibited by the subject when viewed at the correct film speed (18 fps) that is not matched by the exaggerated, "forced, high-speed" human compliant gait referred to by Daegling and Schmitt. This point of film speed was rather significant to Don Grieve, who concluded that, based on the pattern and speed of the subject's gait, the possibility of fakery was virtually ruled out if the film was shot at 16 or 18 fps, rather than 24 fps. The question remains whether a human subject using a forced high-speed compliant gait would exhibit a fluid pattern of joint excursions comparable to those of the film subject. That remains to be demonstrated.

In sum, there has been considerable discussion of various individual aspects of the anatomy, locomotion, and general behavior of the Patterson-Gimlin film subject, with opinions and interpretations sometimes inexplicably varying widely between the experts. Many of these discrepancies are simply repeated without discussion or qualification. For example, Napier expressed disapproval of what he saw as incongruence between the film subject's height and the length of its feet. He determined on the basis of a simple formula (stature = foot length × 6.6) that a foot of between 14 and 15 inches in length indicates a stature of between 7' 8" and 8' 3", a range considerably higher than most estimates. This simplified formula does not account for expected variation present within a human population and it further assumes that a similar relationship holds true for a different and markedly larger species. Curiously, Geoffrey Bourne of Yerkes Primate Center drew a very different conclusion. He said, "It appears to be about seven feet in height, but it left footprints that were rather small for such a big animal—only fourteen and a half inches long." My own study of sasquatch footprints in general, and specifically the film site footprints, suggests a disproportionate length of the toes and heel by comparison to the human foot. Assuming these add approximately 10 percent to the relative length of the foot, an adjusted height estimate according to Napier's *human* formula falls closer to seven feet. Making further

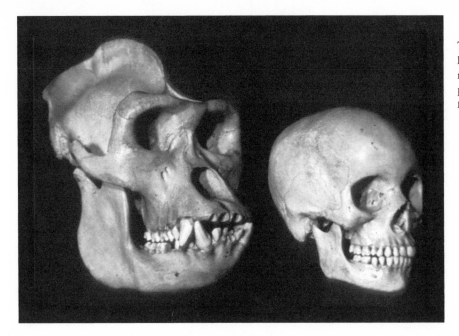

The male gorilla skull on the left displays a prominent bony ridge, the sagittal crest, which provides expanded attachment for the chewing muscles.

allowances for differences in the height of the cranium, the estimate is even more congruent with other height estimates.

This raises the issue of the hominoid crest. Exception has been taken to the combination of an apparent sagittal crest with the presence of what seem to be female breasts. The sagittal crest is a bony ridge that runs along the midline of the skull. It is often characterized as a male feature, but is really a function of the ratio of body mass to cranial volume and the robustness of the jaws. The muscles necessary to operate the jaws and chew a coarse diet require sufficient surface area on the cranium for attachment. If sufficient area is not provided by the rounded contour of the cranium then a projecting flange of bone creates additional surface. In relatively smaller species of primate, such a situation usually only occurs in more robust males who sport larger jaws and more pronounced canines. However, in larger species where the disproportionate increase in muscle mass, i.e., volume, outstrips the increase of bony surface area, larger females may also exhibit sagittal cresting, as is the case with some female gorillas. These are not as pronounced in their development as are the male's crest, but to suggest this is unprecedented reveals a lack of understanding of the biomechanics underlying this trait and an ignorance of the anatomy observable in available museum specimens. In addition, the gorilla's head is capped by a pad of fibrous connective tissue that further adds to the peaked appearance of the head. Given the estimated mass of the film subject, the presence of a crest, even on a female, should come as no surprise. However, it is not at all certain that a bony crest is solely responsible for the subject's peaked head. In an interview with Ivan Sanderson, Patterson noted, "She seemed to have a sort of peak on the back of her head, but whether this was longer hair or not I don't know." Careful observation suggests that a degree of the peaking may be due to a shock of hair that seems to bounce in step on occasion. Therefore, determination of the extent of development of a sagittal crest on the film subject is uncertain.

Colin Groves, a primatologist at the Australian National University, has noted with regard to the film subject that, "There is a big sagittal crest, implying heavy jaws and temporal musculature, yet the jaws are in fact not large or protrusive." Russian investigator Igor Bourtsev drew attention to film frame 344, which depicts a clear lateral view of the facial profile, just before the film subject turns to look at Patterson. Although the pertinent landmarks cannot be precisely discerned, it appears that the head is tipped forward slightly relative to the Frankfurt horizontal, a line tangential to the inferior margin of the orbit and the top of the ear canal. The line is typically horizontal when the head is held at rest, but may be inclined when the head is leaning into a brisk stride. By orienting skulls to this horizontal plane, consistent comparisons can be made. The jaws of the film subject do project beyond the rather flattened nose, but are certainly nowhere near as projecting (prognathic) as a chimp or gorilla. However, flattening of the face is an adaptation exhibited in a number of primates whose diet is coarse, including *Gigantopithecus*. This rearrangement shifts the loading of the bite to the enlarged premolars and molars. The lower face appears quite deep vertically and the lower jaw becomes deeper posteriorly with evidence of flaring of the angle of the mandible. This flaring provides greater surface area and leverage for the enlarged masseter, the principal chewing muscle in front of the ear. This muscle is quite evident in the film subject. Its relative thickness is seen in the pronounced vertical ridge indicating its anterior margin.

That the face and jaws are massive is attested to also by the correlated presence of prominent muscles in the back of the neck and shoulders that attach to the posterior base of the skull and balance it over the spine. These attach relatively higher on the skull than is the case in humans, which accords with the slightly stooped and forward carriage of the head, relatively small braincase, and large jaws. However, the attachment point, is neither as high nor as prominent as that in quadrupedal apes.

Frank Beebe, of the British Columbia Provincial Museum, questioned why a creature with a tall bony crest on its skull has a nonprotuberant abdomen. This objection assumes that a large ape would have a protruding stomach to accommodate the capacious gut required to process a coarse herbaceous diet. Four factors are overlooked in this assessment. First, an assertion of a "tall" sagittal crest is equivocal, as has been discussed.

Second, the reputed diet of the species represented by the film subject is omnivorous rather than strictly herbivorous. The menu of food items reported by eyewitnesses ranges from berries to elk. It is unjustified to assume a diet for sasquatch requiring extensive fermentation of bulk vegetative foods. Kenneth Wylie, a Ph.D. in African Studies, notes that a "flat stomach in an omnivorous or carnivorous terrestrial primate of such bulk would be unprecedented." I must assume that he implies that the existence of an omnivorous or carnivorous terrestrial primate of such bulk would be unprecedented. In fact, we have seen that *Gigantopithecus* has already set that precedent, since microwear analysis of its teeth indicate an omnivorous diet similar to that of a chimpanzee.

Third, if it is implied that sasquatch exhibits an *unreasonably* flat stomach, then an understanding of comparative abdominal anatomy is lacking. A specialized shortening of the lower back and a broad forward-facing pelvic architecture accentuate the protruding gut of the African apes, especially the larger gorilla, whose diet includes more foliage. The film subject has a lower back and pelvic structure more similar to a human, in accordance with its bipedal posture and locomotion. It does not appear to have a shortened lower back and its pelvis appears more basin-shaped than an ape's. Hence, a relatively capacious abdomen would not appear so prominent in a sasquatch torso.

However, in the fourth factor it differs from human torso morphology considerably. In addition to a very wide torso, the film subject has a noticeably barrel-shaped torso, being nearly as deep from front to back as it is wide. Therefore, the volume of the torso is in fact relatively large for its size compared to humans. In all, the generalized anatomy seems consistent with a moderate-length gut combined with a generalized omnivorous diet.

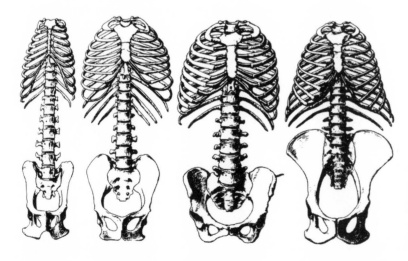

The torso skeletons of a macaque monkey, a gibbon, a human, and a gorilla (Reproduced from A. H. Schultz, 1950, "The Specializations of Man and His Place Among the Catarrhine Primates," Cold Spring Harbor Symposium on Quantitative Biology, Vol. 15, p. 41, Fig. 1.)

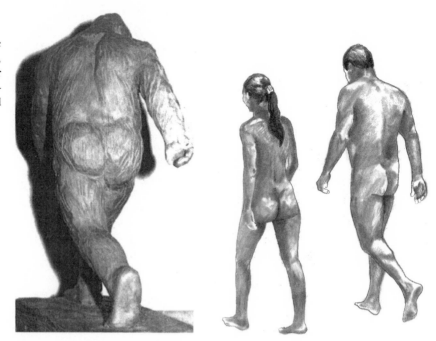

Rear view of a sculpture of the Patterson-Gimlin film subject, (created by and courtesy of Igor Bourtsev) contrasted in approximate scale to a human female and male figure

Napier took exception to the chimerical combination of apelike and humanlike characteristics in the film subject. He remarked upon this after viewing the film and reiterated in his book, "The upper half of the body bears some resemblance to an ape and the lower half is typically human. It is almost impossible to conceive that such structural hybrids could exist in nature. One half of the animal must be artificial." This statement was made in 1972 and although early hominid fossils had been studied by that time, it wasn't until 1974 and the discovery of an *Australopithecus afarensis* skeleton that came to be known as "Lucy" that the mosaic nature of hominid evolution began to be fully appreciated. Lucy exhibited an anatomy in its upper body that was remarkably apelike, while the lower half bore many of the hallmarks of rather humanlike bipedalism. Napier's statement was premature, and in fact the anatomy of the film subject actually anticipated what has become accepted as congruent with the parallel evolution of hominid bipedalism.

Several scientists have made a point of the similarity of the film subject's walk to the walk of a human male, even though the subject bears the pendant breasts of a female. The distinctions between genders in human walking are the result of the wider female pelvis, necessitated by the large size of the birth canal needed to accommodate the human neonatal brain. This adaptation obliges a greater distance between the hip joints in the human female pelvis, which exaggerates the side-to-side tilt of the hips during walking. A bipedal ape of relatively smaller cranial capacity would have no such obstetric constraints and both male and female pelves of that species would be of similar architecture. Therefore, there simply is no incongruence. A "masculine" walk, by human standards, is exactly what should be expected for a small-brained female bipedal ape.

Pendulous breasts, it has been suggested, are a strictly human characteristic and are not to be expected on a creature with such apelike characteristics. The evolution of the human female breast has been a topic of extensive discussion and speculation in the scientific literature. There is no consensus regarding the selection pressures for human breast development or for the timing of their appearance in hominid evolution. In fact, the breasts of *lactating* apes are considerably enlarged, some even falling within the lower range of human dimensional variation, which is notably considerable. However, ape breasts do not have permanent fat deposits as do typical human female breasts, and they are certainly not as prominent as those present on the film subject. Whether the film subject is lactating or not is obviously unknown, however given the apparent engorgement it is tempting to speculate that she may indeed be lactating. Furthermore, if the enlargement of human breasts is somehow linked to bipedalism, as has been hypothesized in at least one scenario among many, then why should it be considered aberrant for another bipedal hominoid to have evolved them convergently?

Concern has been raised over the light color of the soles of the feet, which have been said to be lighter than the palms and much lighter than typically seen in dark-colored great apes. In the film, the soles of the feet are similar in shade to the sandbar, so much so that some have suggested that perhaps sand was adhering to the wet feet of the film subject if it had just crossed the creek. What is more likely is that the smooth surface of the sole of the foot, when angled appropriately to the sun and the camera, was overexposed on the film. When I had the opportunity to visit the film site I was impressed by the dark gray shade of the sandbar. It was much darker than I expected from my viewings of the film. I even scooped some up in a film canister for later reference. Therefore, in the film the sole of the foot appears the same shade as the sand, while in reality the sand is much darker than it appears in the film. Hence, the sole of the foot is actually darker than it appears in the film. The same overexposure also occurs with the palms to a lesser degree due to their different angle to the sun, and is less noticeable due to their limited visibility. They look relatively darker than the overexposed soles, except for occasional highlights. In those frames when the sun does reflect off the appropriately oriented edge of the palm, it appears every bit as light (overexposed) as the soles of the feet. A similar reflective shine can also be seen as highlights on the smooth parts of the face, especially beneath the eye, across the nose, and at the lips. However, due to the more irregular surface of the face, as

A sequence of stills from the Patterson-Gimlin film illustrating the appearance of the sole of the foot (Courtesy of Martin and Erik Dahinden). Below is a sequence of images of a chimpanzee foot seen from a comparable perspective.

compared to the relatively flat plane of the sole of the foot, the overall darkness of the face is more readily perceived. A simple comparison to the appearance of the feet of a chimp or gorilla when viewed from a comparable angle with similar lighting conditions reveals their remarkable similarity.

Frank Beebe was adamant that the film was, as he put it, "a double-jointed phony." He asserted that there was a dead giveaway of fakery in the casts and the photographs of the footprints. "There is plain evidence of a double tarsal joint—a phenomenon that appears nowhere in nature. It can be explained only by picturing a human foot inside some larger mechanical device to make the tracks."

It isn't clear whether Beebe was referring to the so-called split ball feature, which is only mildly expressed in the filmsubject casts, or whether attention had been drawn to the midtarsal pressure ridge. In either case Beebe was apparently unfamiliar with ape feet, which exhibit a much greater degree of flexibility of the midtarsus, called the midtarsal break (see chapter 13). He may also have been unfamiliar with infant human feet, which possess a flexion crease across the ball of the foot very similar to that evident in the casts of the film subject. In persons with hypermobile feet, this crease may persist for some time.

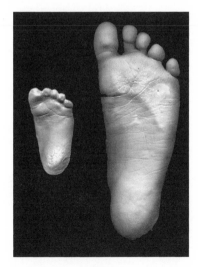

Casts of infant and juvenile human feet exhibiting a flexion crease at the hallucial metatarsophalangeal joint, the "ball" of the foot

I was curious what sort of evaluation the film would receive today from disinterested experts in human locomotion nearly four decades later. I posted a query on the Biomechanics discussion list on the World Wide Web. It seemed reasonable that with the advances in technology a definitive quantitative analysis of the film subject's pattern of movement should be possible. There were three takers of the challenge, but only one ultimately followed through and offered an opinion. Dr. Gordon Robertson of the University of Ottawa, an expert in biomechanics and gait analysis remarked, "My only comment would be that the movements of the subject do not appear to be that of an animal, for example a bear, because of the rather blasé attitude towards the filmmaker. All wild animals would have reacted more cautiously. But as to the subject being a human or another species the film is not of sufficient quality to reach any conclusions. Thus, trying to determine if the gait parameters were similar to a human is impossible. Furthermore, any human can mimic other gait patterns so that discrimination between human and non-human gait is also impossible especially with such a short amount of data."

Robertson's generalization about wild animal behavior seems too simplistic, and the evidence does not seem to justify his characterization of the film subject's "attitude." However his cautionary note about the challenges of distinguishing the film subject's gait from a human's gait is well taken, although perhaps also overstated. It should be noted that this preliminary assessment was based on a dubbed VHS tape several generations removed from the original film.

Advances in graphic technology have made it possible to visualize and study the Patterson-Gimlin film in novel and revealing ways. What lies beneath the hairy covering of the famous California Bigfoot? Animator and computer-generated effects expert Reuben Steindorf, of Vision Realm, created a computer model of "Patty," as the film subject has been nicknamed. Beginning with a reconstruction of the skeletal proportions of the subject's foot that I inferred based on tracks from the film site, combined with casts of the tracks themselves, Steindorf reconstructed Patty's skeletal anatomy from the ground up, using "reverse kinematics." This is a laborious process that creates linkages between the modeled skeleton and the movie image, matching limb segments and joint centers identified on the film subject. Once the superimposed correlation is established, the model can be tracked through three-dimensional space. His digitally animated model comes to life as an animated skeleton tracing Patty's movements across the monitor screen. The model skeleton can be rotated in 3-D space, permitting a view from any desired perspective.

By drawing attention to skeletal landmarks, the stooped posture of the head and shoulders is emphasized as Patty walks smoothly on flexed knees and hips. This is a compliant gait and is adaptive for a heavy biped. A backpacker with a heavy pack

Reuben Steindorf reconstructed the skeleton of the film subject using reverse kinematics. (Courtesy of Doug Hajicek / Whitewolf Entertainment, Inc.)

typically adopts this manner of walking, especially when walking downhill, in order to reduce shock to the ankles, knees, and hips. The compliance of the slightly flexed joints reduces stress on bones and ligaments, and stores kinetic energy in tendons crossing flexed joints.

A consequence of the flexed knee of the support limb is reduced clearance for the swing limb, the limb in forward motion. In other words, the distance between the hip and the ground is less than the length of a fully extended support limb. This necessitates a higher step by the swing limb in order for the toes to clear the ground. Imagine walking with swim fins on one's feet as an extremely exaggerated example of this high-stepping walk. Also noteworthy is the sharp toeing out (abduction) of the foot during the swing phase. Together with the high step, this toeing out may aid in clearance of the swing foot.

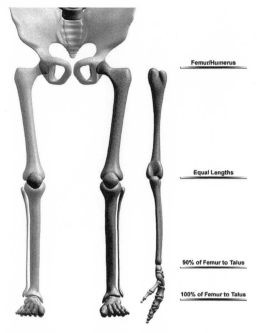

Femur/Humerus

Equal Lengths

90% of Femur to Talus

100% of Femur to Talus

Interlimb proportions of the film subject based on Steindorf's estimation and tracking of joint centers (Courtesy of John Green)

During the animation process Steindorf tracked the joint centers through 116 frames of the film, yielding a reliable estimate of the film subject's limb proportions. The problem of foreshortening is addressed by taking the maximal length of limb segments apparent when the limb lies parallel to the film plane. Steindorf confirmed, as had been previously noted by numerous observers, that the upper extremity was rather long relative to the lower. This ratio can be expressed as an intermembral index (IM). It is defined as the combined length of the humerus and radius, i.e., the distance between shoulder and wrist, divided by the combined length of the femur and tibia, hip to ankle, times 100. The intermembral index for the film subject was approximately 88. I had independently made a preliminary estimate of between 80 and 90, a range that brackets Steindorf's value. An average human IM is 71. Obviously, determining the intermembral index from the film frames is not an exact exercise. By this I simply mean it is not as accurate as putting disarticulated dry bones from a museum drawer on a measuring board. However, it is possible to make a reasonable estimate for the sake of comparison with a would-be human actor. The centers of rotation of the joints can be aproximated for the moving limb segments and proportionalities of the limbs can be determined even in the absence of an absolute scale. Remember an index, being a ratio, has no units. The fact that my own preliminary estimate of IM between 80 and 90 was independently confirmed by Steindorf's more systematic analysis yielding an IM value of approximately 88 should not be casually dismissed. This estimate is well outside the human range as reported in various sources of anthropometric data, so much so that the margin of measurement error in estimating the index for the film subject is trivial. The combination of these proportions with the exceptional breadth dimensions argue compellingly against the simplistic hypothesis of an average man, even one wearing shoulder pads, donning a typical gorilla costume, or using artificial arm extensions.

During the transfer of the film to high definition tape, a pair of unusual details caught Doug Hajicek's attention. There was a peculiar orientation of the swinging leg during the high-stepping walk. The knee appeared to walleye inward during the swing of the limb, while the foot was noticeably abducted or rotated outward. The

A comparison of the effects of scaling the upper and lower extremity of Bob Heironimus to the limb proportions seen in the film subject (Courtesy of Martin and Erik Dahinden and Chris Murphy)

Russian observers had made a passing note of the same thing. In his book *American Bigfoot: Fact or Fiction,* Dmitri Bayanov observed, ". . . Patty left footprints following a single invisible line. As revealed by some of the frames, with each step she made with her lifted foot a kind of rotation movement, as if dancing the Charleston, and put the foot down exactly in front of the one on the ground." Was this a strategy for clearing the foot when walking on a flexed, compliant limb? Or might it indicate an injury resulting in a compromised function of the limb? Obviously, generalizations based on a sample of one should be entertained cautiously.

At a particular point in the film an unusual looking bulge on the right outer thigh caught Hajicek's attention. It protruded noticeably during extension of the knee in support. I discussed possible interpretations of the bulge with Dr. Andrew Nelson, of the Idaho State University health-professions faculty. The bulge appears consistent with a traumatic rupture of the lateral fascia over the thigh. Herniations of muscle through a defect in the overlying fascia of the thigh are rare in humans, but when present in the thigh usually occur over the quadriceps, specifically over the vastus lateralis muscle. The lateral fascia is a layer of fibrous connective tissue that surrounds the muscles of the thigh like a support hose. A rupture may occur directly, as a result of a traumatic blow, or alternately, it may be the indirect result of pathology such as

Dr. Andrew Nelson, Idaho State University, comments on the unlikelihood of a hoaxer in 1967 having an understanding of gait biomechanics. (Courtesy of Doug Hajicek / Whitewolf Entertainment, Inc.)

trichinosis, typhoid fever, or other infections. When the muscle contracts in the early part of the step it protrudes prominently through the gap in the fascia. It reduces when the muscle relaxes. Accordingly, the bulge apparent on the thigh of the Patterson-Gimlin film subject is prominent precisely at the period of the step cycle when the vastus lateralis is contracting and abates when the muscle is relaxed. What are the odds of such details of anatomy and pathology being functionally incorporated into a costume from 1967?

Dr. Nelson concluded, "After analyzing the biomechanical issues of the Patterson footage, I find it very hard to believe that somebody in 1967 could have fabricated the intricacies of the soft-tissue anomaly as evidenced by the irregularity of the thigh. The understanding of biomechanics at that time was primitive at best."

Kyle W. Council, a professional computer graphics animator, offered his perspective of the film. "As part of my craft I have an intuitive sense of natural human movement. I animate human characters for games. When I view the Patterson-Gimlin film, I see a mode of movement that is not human, but humanlike. The arm swing and movement of the shoulder girdles are clearly that of a truly massive frame. The musculature of the creature is very thick, but not restrictive to the range of motion in the shoulder joints. In the trapezius muscle group, the diamond-shaped set of muscles that anchor the shoulders, nape of the neck, and upper thoracic vertebrae, the contraction of these thick muscle sheets is clearly visible. In the cadence of the walk, the presence of great and compact body mass is evident. Also, the creature's knee, not reaching full extension, is certainly an adaptation to supporting great weight. A fully extended knee has poor resistance to twisting force.

"In conclusion, the reality of the film subject as an uncataloged animal is self-evident. It simply cannot be a costume; the boundaries of the human form do not even fit within the form of the creature. Furthermore the mass of the creature is so great and carried with such poise, a man could not even be trained to carry such great weight or walk bearing weight in such a fashion. The muscle masses, if they had been padding, would have been static and restrictive. The presence of visible muscle contraction, the freedom of movement, and the cadence of the walk, all fully support the claim that the footage documents a living hominid outside the genus of man."

I found Council's comment that ". . . the boundaries of the human form do not even fit within the form of the creature," a significant one. Recall the stark contrast evident between McClarin's profile and that of Patty, as well as the revelations of the *Anthropometric Source Book*. In this vein, the reaction of an avid bodybuilder provided some particularly insightful perspective on the physique of the Patterson-Gimlin film subject. First, he noticed the very large and well separated deltoid muscles. He noted that he had only seen one human being with deltoids that large. It was none other than Arnold Schwarzenegger at his peak. Next, the trapezius attached rather high on the back of the head. He compared them to Dorian Yates and Greg Kovacs, who are noted for their trapezius development; neither had traps remotely as large as seen on the film subject. The appearance of the latissimus dorsi was very unusual in that the large muscle appeared to attach higher up on the thorax than in humans. He observed that the gluteus maximus appeared nearly three times more massive than typical in human athletes. The gastrocnemius, especially the lateral head, and the soleus, comprising the calf muscles, were exceptionally large and well developed. He compared them to those of Mike Matarazzo, known for the largest calf muscles in history—the film subject won hands down. He attested that no bodybuilder, not even Schwarzenegger, came close to the size of the erector spinae muscles, the two columns of muscle running parallel to the backbone. Finally, the complete naturalness of appearance and movement of the scapulae, or shoulder blades, impressed him.

An enlightening comparison can be made between the dimensions of Arnold Schwarzenegger at the peak of his bodybuilding career and estimated dimensions provided by Chris Murphy, based on a reconstructed height at the upper range of estimates for the Patterson-Gimlin film subject:

IN/LBS	SCHWARZENEGGER	P-G FILM SUBJECT
height	74	88
weight	235	700
arms	22	32
chest	57	68
waist	34	68
thighs	28.5	46
calves	20	30

There is certainly a diversity of perspectives and conclusions about the nature and implications of the Patterson-Gimlin film. Those that have given the film the most thorough consideration have, with few exceptions, concluded that it is probably authentic or that its authenticity cannot be readily eliminated given the limitations of image quality or want of definite scale. It should be remembered that the original film was viewed a number of times before it was duplicated, rendering it somewhat scratched and damaged. The images that most people see today are the product of several generations of copying from film to film and then to video or digital image. This has drastically reduced the clarity and detail of the image and potentially introduced artifacts that might distract the casual observer from the straightforward impression of the image. On the other hand, those who are most adamant in their rejection of the film, characterizing it as an obvious "man in a monkey suit," are often the least familiar with its details and its provenance, or they have little or no expertise with which to assess it meaningfully. After nearly four decades, the film remains a most intriguing piece of natural history footage.

9

Ape Antics: Behavioral Parallels

Bipedalism in primates is a behavior that has typically been associated strictly with human evolution. The assumed singularity of this adaptation has been recently challenged by the suggestion that other ape species, e.g., *Oreopithecus,* independently evolved bipedalism. Furthermore a growing and diverse hominid fossil record has revealed a number of apparent strategies for practicing bipedalism among direct human ancestors and closely related collateral hominid lineages. Still, the habit of walking on two legs has remained the hallmark of differentiation between our hominid ancestors and the closely related antecedents of the surviving great apes. On this basis some have suggested that as an upright bipedal primate, sasquatch must represent some form of "missing link"—an early branch of the human family tree. They propose that if it exists, sasquatch is a relic hominid, such as *Paranthropus,* or an even later hominid such as *Homo erectus* or *Homo neandertalensis,* rather than a relic Pleistocene ape such as *Gigantopithecus.* Hominid status would imply that sasquatch are decidedly more humanlike than apelike, not only in its anatomy but in its behavior and intelligence as well.

Generally, when two closely related species share a feature, such as bipedalism, the simplest or most parsimonious explanation for the shared anatomical/behavioral adaptation is that it arose only once in the past in a common ancestor and was subsequently inherited by both descendant species. In this case it would be argued that bipedalism, in both humans and sasquatch, would have been inherited from a common ancestral hominid species. This certainly remains a possibility worthy of consideration; however, it has become very clear that examples of multiple origins of anatomical and behavioral traits, even some seemingly quite complex yet remarkably similar, are rampant in evolutionary history. Nature frequently solves the common challenges of life by arriving at remarkably analogous solutions, even from rather disparate starting points. For example, biologists have concluded that eyes have evolved independently as many as forty times. By comparison, it seems a relatively simple matter for a terrestrial ape, such as *Gigantopithecus,* to have evolved bipedalism independently, yet convergently with early hominids. When faced with moving between patchy resources, the same type of environmental challenges that exerted selection pressure on early hominids in Africa and/or southwest Asia could well also have influenced the evolution of particular apes in Eastern Asia. This seems all the more likely when the potential effects of large body size and the exploitation of more temperate habitats is taken into account. The energetics of walking efficiently on two limbs instead of four would have similarly influenced other aspects of anatomy such as limb proportions, further adding to the superficial resemblance of sasquatch to the human body form. The Patterson-Gimlin film subject's intermembral index has been estimated to be approximately 88, meaning the arms are slightly shorter than the legs. This value is intermediate to the average indices for gorilla (116) and human (71) and indicates an emphasis on hindlimb-dominated locomotion with elongated legs for efficient walking and shortened arms for more economical arm swing.

More compelling than the convergent appearance of upright posture would be aspects of behavior that set the members of the genus *Homo* apart from the apes. How do these activities compare with behaviors reported in sasquatch encounters? The archaeological record provides the hard evidence for some aspects of behavior. The artifacts associated with hominid evolution reveal the use of simple stone tools by as early as 2.5 million years ago, with evidence of modified digging sticks employed even earlier. By 1-2 million years ago, distinctive and sophisticated stone-tool cultures are associated with hominids that were systematically hunting big game and returning to well-established home bases where there was evidence of controlled fire use. Endocasts of the hominid braincase reveal asymmetries that indicate the development of the language centers of the brain by this period as well. Indications of abstract thought are represented in aesthetic objects of art and later through self-adornment.

In stark contrast, eyewitness accounts of sasquatch are noticeably devoid of such behavioral and cultural elements. There is a consistent lack of observed tool use, home bases, permanent shelter, controlled fire, clothing, language, art, or gregarious social structure beyond the mere rudiments of these behaviors, to the extent that such have been observed and documented among the known great apes. Sasquatch *could* be an early offshoot of the hominid radiation, yet lack indications of the behavioral aspects of later hominid development, or instead, it might be descended from *Gigantopithecus,* or another Pleistocene great ape, perhaps a distant bipedal cousin to the orangutan. In the later case, its superficial physical resemblances to human posture and gait would be the result of convergent evolution of a locomotor anatomy similar to hominids, selected for under similar ecological conditions.

North Americans are accustomed to seeing most familiar animals, wild or domestic, postured horizontally, on all fours, not standing upright. Even a bear, when seen upright, conveys a decidedly humanlike impression, earning it a peculiar notoriety in folklore throughout history. When Europeans were first exposed to specimens of great apes, they marveled at their remarkable similarities to humans. Those similarities compelled Linnaeus, the founder of modern systematics, to undertake the bold step, in 1735, of including humans as a species of primate—the "first or top ones." He christened man as *Homo sapiens* (the *wise* man) acknowledging the human's singular intelligence, which separated him from the remnant of the animal kingdom, but confessed that he could find little physical difference between man and ape. "It is remarkable," he concluded, after carefully comparing their anatomies, "that the stupidest ape differs so little from the wisest man, that the surveyor of nature has yet to be found who can draw the line between them." Still, man's upright posture was taken as a symbol of his divine dominion over all of creation.

One notable distinction in behavior was the ape's awkwardness when attempting to stand upright. This handicap was frequently depicted graphically at the time by portraying the ape supporting itself with a walking stick or staff. It should not be surprising then that an enigmatic ape that walks effectively upright like a human would give an even more dramatic impression of its apparent "humanness," even if the remainder of its behavior offers little evidence to support that initial impression. We have pointed out the general absence of behaviors in sasquatch that are associated with hominid evolution, but what specific details of sasquatch behaviors suggest similarities to the living apes?

Dr. John Bindernagel, wildlife ecologist in British Columbia, drawing upon a steady stream of eyewitness reports from western Canada, many of which he has personally investigated, is impressed by the notable consistency in subtle behavioral and anatomical details mentioned by observant eyewitnesses after encounters with sasquatch. In his book, *North America's Great Ape: the*

Early depiction of a specimen of a chimpanzee described by Tyson in 1699

Dr. John Bindernagel, British Columbia, has pointed out the numerous parallels between reported sasquatch anatomy and behavior and that of known great apes. (Courtesy of John Bindernagel)

Sasquatch, he points out significant and compelling parallels between these details of reported sasquatch activities and what has become generally known only recently about the behavior of the great apes. For that matter, Native American traditional knowledge of wildmen appears to be based on the observation of natural behaviors that are echoed in contemporary accounts of sasquatch by Anglos, and find their equivalents in recent field observations of known Old World apes.

For example, on several occasions witnesses have reported that the sasquatch smiled or grimaced at them. This facial expression appears to have been captured by the artists that rendered the Tsimshian monkey mask and the stone heads from the Columbia River gorge. It is quite vividly represented in a more stylized form in the *buk'wus* masks of the Kwikiutl tribe. In primates, and especially the apes, the expanded muscles of facial expression have become much more elaborated, permitting more complex combinations of lip and brow positions to convey socially appropriate meanings. The smile, or grimace, is a complex facial expression in primates that taken in context conveys emotions ranging from disgust and fear to appeasement and happiness. In apes, when two animals meet, either or both may grin, the subordinate out of fear or submissiveness, the dominant as a gesture of reassurance. Most mammals common to North America do not possess such an extensive repertoire of facial expression and that limitation is usually evident in their stoic representations in totem art and ceremonial masks. In contrast, a distinctly apish quality of facial expressivity seems to distinguish depictions of the wildman of the woods.

The female counterpart to the *buk'wus* is the *dsonoqua,* which adorns many totem poles, clan crests, and ceremonial masks of northwest Native Americans. She is depicted with distinctly puckered or trumpet-shaped lips, which calls to mind the remarkably similar facial expression of a hooting chimpanzee. Here again it is the representation of the expressive muscular lips of an apish mouth that sets off this stereotyped representation. The classic account by William Roe in 1950 relates the sasquatch pulling branches to its mouth and dexterously manipulating and stripping the leaves with its lips, further reminiscent of the muscular muzzle of an ape employed in food processing, as well as in communication through facial expressions. Indeed it is interesting to note that the repertoire of facial expressions utilized in chimpanzee communications has been codified into a number of emotions, and they all principally involve mouth postures with various degrees of baring teeth or contorting lips. These are precisely the two principal themes prominently embodied in Native American representations of the *buk'wus* and the *dsonoqua*.

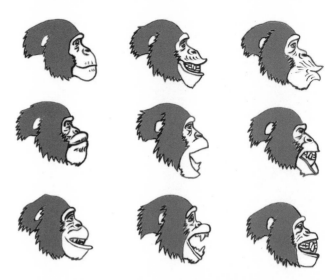

A representation of typical chimpanzee facial expressions. Note the emphasis on lip position and baring of teeth.

Emotional expression is not only limited to facial displays. The hair of a sasquatch, especially over the head and shoulders, has been reported to occasionally bristle or stand on end, an action that is called piloerection. One particular witness to an encounter with a sasquatch noted the hair rise and fall repeatedly as the sasquatch stared at him. Dr. Jane Goodall reports of the Gombe chimps, ". . . a chimpanzee bristles when he is highly aggressive, or socially excited, or when he sees or hears something strange and frightening . . . Sometimes his hair rises and falls alternately, as, presumably he reacts emotionally to what he hears or sees. Subordinates often become nervous and move away when nearby superiors suddenly show hair erection; a bristling male is more likely to display

A chimpanzee illustrating the trumpeting of muscular lips in a hoot call (Courtesy of Alain Houle). Notice the striking similarity to the posture of the lips in the *dsonoqua* masks and totem carvings.

An agitated male chimpanzee displaying with hair standing on end (piloerection) (Courtesy of Jack Murphy)

or attack than one who is sleek." The hair-raising reaction of the sasquatch to an encounter with a human witness is quite consistent with this ape behavior.

The dexterity of apes lends them the ability to manipulate their environment to a greater degree than many animals. Among the earliest reports of sasquatch are accounts of stone-throwing, most probably as a form of intimidation behavior. Bindernagel notes, "As early as 1846, stone-throwing by sasquatch was reported in British Columbia. I have since discovered that's a very common element of chimpanzee intimidation behavior, which was really not reported until the 1960s. That suggests a similarity between sasquatch and chimpanzees and a possible relationship between these two animals." Stone-throwing in chimpanzees is now well documented through extensive field observation.

I have personally experienced this in the Siskiyou Mountains of northern California, when professional guide Mark Slack and I were examining a set of very fresh 16-inch tracks crossing the duff-covered trail we were on. A softball-sized stone was apparently lobbed in an arc onto the trail a few feet from us. There were no nearby elevations from which the rock might have coincidentally tumbled. By all appearances it was deliberately thrown in our direction.

Sometimes the missiles are less weighty. John Mionczynski, a Wyoming wildlife biologist, had his tent collapsed at night by a tall upright figure with a distinct "five-fingered hand" silhouetted by the moonlight. After scrambling out and stoking the fire he found he was the target of pinecones that were lobbed at him from the darkness throughout the remainder of the night. This behavior and others such as chest-beating, tree-shaking or striking, and loud vocalization by sasquatch compare with similar antics in great apes. They are characteristic of ape intimidation behavior.

Another type of intimidation behavior, which has been described for sasquatch, is pushing over dead snags in the direction of the intruder. Woodcutters in southeastern

Idaho reported a definite reaction to their activities, including loudly broken limbs and deep guttural growling vocalizations that culminated in large snags being pushed over in their direction. At this they terminated their woodcutting and withdrew from the area. Large footprints were later found in the vicinity. Biruté Galdikas reported similar behavior in orangutans. During her first encounter with an adult female orangutan, she heard cracking noises overhead and looked up to see the orangutan very deliberately pushing a dead snag toward her.

The sound of striking wood on wood has frequently been associated with reported sasquatch activity, although I am not aware of any *direct* observation of sasquatch engaging in this behavior. Bob Titmus recounted an experience that occurred in 1959 when tracking a sasquatch on upper Bluff Creek together with Art Long. The tracks they were following were extremely fresh. Titmus recalled, "Shortly after they [the tracks] left the creek and started climbing the mountain, we were brought up short by what sounded like something pounding on a rather hollow tree or perhaps log with a very big club. These sounds were being made in what could be taken to be a coded series. Eventually I found a large stick and a log and started pounding out a series of my own. Each time I stopped my series of pounding, it was repeated from above exactly, both in timing and the number of blows I had struck, only very much louder. This continued for some fifteen or twenty minutes . . . I am just as convinced today as Art and I were the day it happened that one of the Bigfoot creatures was doing the pounding some thirty or forty yards ahead of us in the dense timber and undergrowth. Art refused to move one step further in that direction and wanted only to return posthaste."

Drumming by chimpanzees may take the form of excitement behavior or as a form of long-distance communication. Gorillas employ a form of ground-slapping in relation to group movements. When a human observer agitated her, a female gorilla clapped her hands to attract the attention of a silverback at a distance away. The slapping of the ground, trees, or the chest, is also a form of intimidation behavior. Primatologist Chris Boesch reported that the chimp study group would split during bouts of extended foraging, but would drum to announce their positions and maintain aural contact. It was possible to track the group's movement by the orientation and distance between trees drummed upon in sequence. Likewise, Clayton Mack, a Native American from Bella Coola, B.C., concluded much earlier that sasquatch communicated their location to one another by striking tree trunks. By noting the direction and interval of the percussion their location could be determined.

The construction of crude sleeping nests, from grass, ferns, leafy vegetation, or boughs has been attributed to sasquatch. Professional field geologists discovered large

footprints in a remote area of northern California associated with a simple plaited bark pad or pillow stuffed with ferns gathered from the surroundings. They had first noticed that all the ferns were curiously pinched off at ground level in an area, and then found the pad beside a large tree with signs of rubbing against the trunk. They surmised that it was either a vantage point to observe the approach from below, or perhaps a spot to sun. Only upon further searching did they discover the oversized footprints that would seem to associate the activity with a sasquatch.

Wildlife biologist John Mionczynski and I have observed a large bed of crudely interwoven pine boughs, broken off and gathered from some distance, while in the Blue Mountains of southeastern Washington. The great apes likewise construct sleeping nests of interwoven vegetation, either in the trees, or in the case of the large male gorillas, on the ground due to their great body weight.

Frequently, it appears that the sasquatch may simply bed down in the open. A witness to a pair of female sasquatch sleeping in the open described their posture as lying face down with their arms and legs drawn in and tucked under them. The sleeping posture of gorillas has also been described in remarkably similar terms.

Occasionally, a distasteful pungent odor is experienced in association with a sasquatch encounter. The odor can be rather overpowering and is compared to the smell of rotten eggs, putrid meat, or rank body odor. However, much more frequently, no noticeable odor is detected during an encounter, even at close quarters. A mere 10 percent of the reports accumulated by John Green make any mention of an odor. In his interactions with the mountain gorilla, Dr. Schaller noted an odor described like pungent human sweat, manure, and distant burning rubber. He suspected it emanated primarily from the silverbacks when the group was in a state of excitement. Indeed, the male gorillas have well-developed axillary organs, located in the armpits, comprised of apocrine sweat glands. The same type of glands develop to a lesser degree in humans with the onset of puberty. These can reflexively discharge a strong musky odor in response to fear or threat. Dian Fossey recounted one of her early encounters with a charging silverback gorilla when she approached the group too closely. The onrushing patriarch gorilla stopped just short of her position, but she was hit by a powerful musky odor that emanated from the ape. This function of well-developed ape axillary organs may explain the inconsistent reporting of an associated strong odor during sasquatch encounters.

The anatomy of the male primate genitalia is often very diagnostic in species identification. Perhaps due to relatively small size or the presence of obscuring body hair, the genitalia of sasquatch are rarely if ever mentioned. This fact is rather odd, given Hollywood's past exploitation of the image of the ape as the embodiment of all of man's baser and unbridled drives and appetites. There is a little known detail of an account rendered by Albert Ostman, a Canadian prospector, who claimed to have been abducted and spent a period of confinement with a "family" of sasquatch in an apparent semipermanent camp in the mountains of British Columbia. Under questioning by Dr. Daris Swindler, a primate anatomist, Ostman described his recollection of the adult male's genitalia. He described them as resembling a stallion's, in that the penis was ensheathed. Indeed, the great ape penis consists of two components, a hair-covered *pars proximalis* that ensheaths a hairless or glabrous *pars distalis,* which in turn extends from its sheath upon erection. The shaft in chimpanzees is typically sickle-shaped and ends in a taper. It lacks a differentiated glans capping the shaft, in contrast to the human organ. When flaccid, the ensheathed penis is quite unobtrusive, especially when obscured by a profusion of body hair.

All of these behavioral and anatomical parallels, however, beg the question of how a great ape would make a living in temperate Pacific Northwest forests. It has been argued that apes are strictly tropical primates that require a constant food supply to nurture the development of large brains and therefore are limited in distribution to lush equatorial tropical forests. This generalization is drawn based on the limited sample of modern apes inhabiting tropical refugia—a mere relic of the taxonomic and ecological diversity that existed in earlier geologic times. Nor does it recognize the diversity of habitat and dietary preferences that are present even within the limited extant apes of the Old World. Some of these habitats show marked seasonality in food availability, rainfall, and temperature. These variations may not be as pronounced as those present in many temperate forests of western North America, but to suggest that apes rely on a uniform and constant habitat is an oversimplification, to say the least. More fundamentally, it ignores the evolutionary history of the apes, which the fossil record indicates took place in large measure in temperate and subtropical forests in Eurasia.

Based on the numerous accounts that mention feeding or carrying food, the sasquatch diet seems to span the wide spectrum of a generalized omnivore. Eyewitnesses have reported everything from roots and berries to deer and elk. The latter would seem to be out of step with the commonly stereotyped image of the vegetarian ape. Eating meat, let alone preying on large mammals, seems out of character for a great ape, but numerous credible eyewitnesses recount seeing sasquatch dispatch and carry off adult deer as well as fawns. Recently, the role of meat eating in apes' diets, especially the chimpanzee diet, has been recognized and more fully documented. Animal protein in the form of insects, reptiles, fledgling birds, bush pigs, and monkeys, makes up a significant fraction of the chimpanzee diet. These are not merely opportunistic events. The hunting of monkeys is a very intentional and coordinated effort with a remarkably high rate of success.

Descriptions of sasquatch seem to lack mention of a noticeably protruding belly characteristic of a dedicated herbivore. A long and capacious intestinal track is needed for the lengthy fermentation and digestion of a fibrous leafy diet. This anatomy is exaggerated in the apes, particularly the gorilla, due to their funnel-shaped thorax, their pelvis with narrow outlet, and tall broad anterior-facing hipbones. The broad shoulders, deep torso, and short flaring pelvis of a giant biped would accommodate a large abdomen less noticeably. Furthermore, the disproportions of increased body size would mean that abdominal volume increases disproportionately to more apparent linear body dimensions. Witnesses do frequently note that the sasquatch torso is rather cylindrical, being as deep from front to back as it is wide. Krantz noted that the body type exhibited by the Patterson-Gimlin film subject exceeded the proportions of even an extreme endomorphic human physique. These proportions would readily accommodate the gut required for digestion of the diet of a wide-ranging omnivore, such as that inferred for sasquatch.

There is an abundance and variety of wild plant foods to be found in the forests of the Pacific Northwest, which few people are aware of. These resources have sustained two large omnivores, Native Americans and bears. The Plains grizzly first encountered by Lewis and Clark, but possibly sighted recently in Yellowstone, attained a weight of 1,500 pounds. Grizzlies were once common throughout the western part of North

America. Today, it is estimated that there are over 32,000 grizzlies, 22,000 in Canada, but less than 1,000 in the lower forty-eight states, p_ tana, Idaho, and Wyoming. The amount of territory needed to sustain a considerably depending on season, gender, and age of the animal and a food. In Yellowstone National Park, a male grizzly may require up to 10_ miles. Two females had home ranges of 106 and a mere 27 square miles resp_ In the Yukon, ranges average between 28 and 33 square miles; in Alberta, 70 _ square miles. By comparison, it has been suggested, based on pattern analysis _r sasquatch sightings, that a sasquatch home range is under 18 miles in radius, or about 1000 square miles (although perhaps capable of traveling over considerably larger areas to disperse).

Granted, bears hibernate to avoid the privations of winter, and indigenous human populations of the past reduced physical activity during lean periods and relied upon stored foodstuffs to see them through the winter months. The apes have the ability to put on fat in times of plenty in a pattern very similar to humans, and captive apes occasionally suffer from obesity. The fat-tailed dwarf lemur is the only primate known to truly hibernate, and does so during Madagascar's dry season when it survives off the stores of fat accumulated in its tail. The possibility of hibernation or periods of inactivity is one that may have been realized in sasquatch evolution. It has even been suggested that humans show the residual effects of a past history of hibernation in the form of seasonal affective disorder (SAD). Individuals who suffer from SAD have been shown to increase their production of melatonin with changes in photoperiod. Melatonin plays a critical role in seasonal regulation of body temperature and in adjustments of biological clocks. Humans with elevated melatonin experience lethargy, hypersomnia, hypometabolism, increased appetite, and weight gain. This may be a past adaptive evolutionary mechanism of human populations in temperate latitudes.

Patterns of torpor and hibernation are diverse and plentiful in mammals and some birds, prompting the hypothesis that the ability to hibernate is a primitive mammalian condition. Indications are that the use of torpor and hibernation by mammals in the wild is much more widespread than presently appreciated. Research supports the hypothesis that torpor is an evolutionary extension of mammalian sleep, which evolved as means of conserving energy in endotherms, or warm-blooded animals. When energy stores decline, energy is conserved by lowering body temperature during sleep or by increasing the daily duration of sleep. Whether hibernation is ancestral or newly developed, the genes required for hibernation are widespread among mammals. Recently, geneticists have announced the discovery of these genes in the human genome. This opens up the prospect of developing a means to place astronauts in a

Large tracks photographed after a sighting the preceding evening on the Oregon coast, south of Tillamook (Courtesy of the International Society of Cryptozoology)

state of torpor during extended space flights. It also suggests that natural selection for torpor or hibernation in a large primate inhabiting northern latitudes may not be so far-fetched a proposition.

Of course, regional availability of food in winter is a differential challenge depending on the geography, temperature, and the degree of snow accumulation. In the moist and moderate climate of the Pacific Northwest, especially the coastal forests, winter's challenges are less stringent than those potentially encountered at higher elevations in the interior of the Rocky Mountains. There is simply not enough information about the purported diet of sasquatch to make such hypothetical statements about the limitation of critical resources available to them. Nor is there a realistic appreciation by most critics regarding the foods available throughout the year. A Hoopa medicine woman, in northern California, informed me that many traditional food resources actually became more abundant in their forests during the wintertime. Often overlooked, edible tree lichens may represent a large potential source of winter carbohydrate within sasquatch's suspected geographic range.

Concerning potential sasquatch habitat and feeding, Dr. Bindernagel observed, "Many sasquatch sightings come from clam beaches. West coast clam beaches are incredibly rich in terms of shellfish, marine worms, and other forms of animal protein. I think one day we will come to accept that the west coast beaches are one of the best sasquatch habitats in North America." A possible example of a sasquatch foraging beach resources was witnessed on a southern Oregon beach by a couple enjoying a moonlit evening near a large tide pool. They noticed an exceptionally tall and dark figure striding up the beach apparently investigating the margin of the pool. As it drew near they realized it stood over 8 feet tall and estimated that it weighed in excess of 600 lbs. Its stride was exceptionally fluid and rapid. It became evident that it wore no clothing, but was covered with 3–4-inch-long dark hair. When the figure was within 100 feet of the witnesses' position, they retreated up the adjacent bluff and took refuge in their vehicle until it had departed the area. In the morning enormous footprints remained in the dry sand above the high-tide line. The stride averaged 65–70 inches.

In fact, most sightings come from areas that are biologically rich and have all the requirements to sustain the life of a large mammal, such as ample rainfall, lush vegetation, and prolific sources of animal protein and winter forage. John Mionczynski and I have surveyed these temperate "rain forests" in areas of both the Pacific Northwest and the Intermountain West with consistent historical and contemporary reports of tracks and sightings. Without a doubt, there exist ample resources to support a large omnivorous ape in appropriate habitat types, even in the interior Rocky Mountain regions.

The apparent nocturnal activity pattern suggested for the sasquatch has been seen as an unlikely behavior for a higher primate, all of which are active during daylight, or diurnal, and lack the specializations of the retina for night vision. However, recent observations have emphasized the behavioral plasticity of the great apes. Owen Caddy is a professional park ranger with extensive primate study experience in Africa. As a warden for the Uganda Wildlife Authority, in the Murchison Falls National Park, one of his principal duties in the remote Rabongo Forest was to supervise the study of the easternmost surviving populations of chimps in Africa. He also spent time working with mountain gorillas in the south of Uganda. He discovered an unexpected aspect of the chimps' activity pattern. "The Rabongo chimp population is unique in that after years of civil war and poaching, they developed the only documented transition to nocturnal foraging and travel. They go to great lengths to avoid detection, much more so than other chimpanzee populations." It seems that much of the daylight was spent in secluded nesting sites. Foraging and travel took place after dark and in the predawn hours. The chimps seemed to negotiate their surroundings quite effectively, even on moonless nights. Caddy continued, "Nevertheless, after years of work, we were able to develop a methodology to track them, map their travel routes and times, and observe and study them." The potential behavioral parallels between the Rabongo chimps and a "clever North American primate that does not wish to be found" struck Caddy. It

Only in the last several decades have the pioneering efforts of field primatologists given us a truer understanding of the natural behavior of the great apes. (Courtesy of Diane Doran)

would seem that to reject outright the possibility of nocturnal activity in a large primate, even an opportunistic, or facultative nocturnality, as displayed by the Rabongo chimps, is premature.

Ideas about great apes are very different today than were those of the rather recent past, and may continue to be revised as their behavior is better understood. Eighteenth- and nineteenth-century Europeans considered apes the embodiment of all of humanity's baser instincts. Apes were viewed as lustful and savage. Well into the 1930s Barnum and Bailey billed a gorilla named Gargantua the Great as *the world's most terrifying living creature.* A generation of audiences was raised on Hollywood's "King Kong" stereotype of ape behavior. The bestial ape abducting the hapless heroine echoed attitudes with deep historical roots, which were rehearsed in an extensive genre of horror films. The King Kong monster and its various spin-offs were second in popularity only to Frankenstein's lightning-reanimated creation.

In light of this history, there is, with rare exception, a curious absence of such "monster-image" embellishments in the descriptions of encounters with sasquatch, the "giant-hairy-ape." This absence of such trappings would seem rather uncharacteristic, if this phenomenon were simply the metaphorical or mythological expression of man's baser, more brutish, counterpart. Indeed, it seems that many aspects of the traditional Native American perception of the wildman of the woods actually anticipated by centuries, and the historical Anglo perception of the sasquatch anticipated by several decades, a more accurate image of the great ape that prevails today. Now ecotourists vie for the privilege of approaching gorillas in their natural habitat and dedicated field primatologists and conservationists endeavor to learn more about the natural history of our nearest relatives, while struggling to prevent their ultimate demise by extinction. In light of this novel and growing understanding of apedom, it appears that the inferred sasquatch anatomical and behavioral profile is indeed a very real approximation of a North American great ape.

10

SOUND OFF: VOCALIZATIONS

Among wildlife species that are relatively solitary, secretive, and widely dispersed, vocalizations play a particularly important role in their natural behavior. Experienced outdoorsmen and naturalists alike rely on vocalizations in locating and identifying reclusive animals. Over the decades a number of inexplicable cries, whistles, and rumblings have been heard in the mountain forests of North America. Witnesses are not only impressed by their inability to identify these sounds, or to attribute them to a known animal, but they are also awed by the immense volume of vocalizations that indicate an animal of gigantic proportions. A few examples of these unidentified cries have been recorded. Are these calls the voice of an unknown primate?

The vocal repertoire attributed to the sasquatch is extensive and varied. It includes loud screams, wails, and whoops, as wells as whistles, grunts, moans, snarls, and "tooth popping." The most commonly noted sounds are the intense screams that carry over considerable distances, since they are the least likely to be rationalized as coming from a familiar animal or from another person. The loud scream was described by one witness as "an unusual sound of tremendous power coming across the lake. It started as a low-pitched raspy grunt and gradually rose to a high clearer pitch, lasting for eight to ten seconds." I have heard the trailing end of one such call when camping in the Siskiyous of northern California. My companion awakened me at about 2:00 A.M., just in time to catch the sound of a lingering high-pitched wail from some removed location. It would be later that night that our camp was visited by a nocturnal intruder that circled our site with heavy brush-snapping footsteps. At one point it seemed to clack its teeth in rapid series, only to receive a similar reply from the opposite perimeter of our camp. Tooth clacking is a behavior common to anxious primates. In particular, frightened orangutans have been reported to audibly grind their teeth. Also called "tooth popping," this behavior is common not only to primates, but it is also associated with other types of wildlife, such as bear. The particular circumstances of this nocturnal visit seemed to convincingly rule out typical bear activity.

Orangutans are quite solitary by comparison to the gregarious chimps and gorillas. Dominant male orangs vociferously defend large home ranges of up to 12 km² with a booming call, backed up with large body size. The much smaller females travel over a smaller home range, generally encumbered by infants and juvenile tag-alongs. Youngsters stay close to their mothers for five to six years. In order to repulse competing males and to attract sexually receptive females, the dominant male orangs employ their imposing long call. Dr. Biruté Galdikas described it as "the most impressive and intimidating sound to be heard in the Kalimantan forest." The call begins as a slow grumble with vibrato, increases to a leonine roar, then terminates decrescendo with soft grumbles and sighs. They are audible up to 2 km away through the dense Borneo and Sumatran forests. The generally solitary nature and social structure of the orangutan has been compared to that of sasquatch, and the screams attributed to sasquatch may serve the same purpose as the orang long call.

More often than not, the recordings alleged to be of sasquatch origin have been made independent of any visual contact with the vocalizer. The classification of such anonymous vocalizations is relatively subjective, unless quality recordings can be quantitatively analyzed. Dr. Robert Benson, director of the Texas A&M Corpus Christi Center for Bioacoustics, has analyzed a recording said to be a sasquatch sounding off.

Dr. Robert Benson, director of the Texas A&M Corpus Christi Center for Bioacoustics, has analyzed a recording said to be a sasquatch sounding off. (Courtesy of Doug Hajicek / Whitewolf Entertainment, Inc.)

An amateur recording was made near Puyallup, east of Tacoma, Washington, in 1972. The recording was digitized to facilitate computer enhancement and analysis. As he played back the recording, Benson said, "Look at the screen. This is really interesting. See the formant structure? I was hoping for that because that gives us the chance to really make some measurements that can be used to get some idea of what animal may have made these sounds."

When a sound is generated in the vocal tract, the shapes of the pharynx, teeth, tongue, and lips in turn filter it. The resulting concert of frequencies is referred to as the formant structure. Animal vocalizations exhibit a typical identifying formant structure. Benson continued,

"One of the first steps in doing an analysis like this is to see what can be ruled out of the possibilities. You can do this based on your experience. I think that based on the experience I have had, we can rule out barred owl, elk, wolf, and coyote. With some analysis we could rule out some animals not native to the U.S."

Using his trained and experienced ear in concert with modern computer analysis equipment, Benson set out to eliminate the possibilities from among North America's known wildlife by comparing sonagraphic tracings of their calls. He found in this instance that the sounds are likely from a primate, but can they be distinguished from known species, or a human voice for that matter? Benson concludes, "We've been able to compare this sound to the obvious animals that it might be. That includes gorilla, chimpanzee, howler monkey, elk, and wolf. In our analysis so far, it does not seem that the source fits into these groups. However, it does appear from the work we did comparing it to a human voice, that it is probably primate."

Dr. Greg Bambenek, a clinical psychologist and experienced outdoorsman, was witness to a scream that overwhelmed his skepticism regarding the sasquatch. It happened during the Bigfoot Field Researchers Organization (BFRO) investigation at Skookum Meadows in September 2000. Recorded calls were broadcast during the night at half-hour intervals from a hilltop above the base camp, but without response. The next night the system was taken down and returned to the base camp where broadcasting resumed. Dr. Bambenek, accompanied by Allen Terry, returned to the hilltop location after dark. The broadcasts continued at regular intervals. Then at 1:00 A.M. something happened. Dr. Bambenek relates, "I was standing at the rear of the vehicle, which was facing east, and Allen Terry was in front of the vehicle facing west. A very loud scream, very similar to the one we had been broadcasting, came from directly east. There was a three-second interval and a second scream was heard and then a third repetition that attenuated midway, possibly because a pack of coyotes on the facing mountain started howling. This scream was like nothing I have ever heard. I estimated it to be about 100 yards away, but its volume, timbre, and presence was unbelievable. Within the scream I could hear the tympani of a large volume of air being released to manufacture this scream. The quality of this sound had to have been made by an animal with very large lungs, much larger than human lungs and vocal cords. It filled the woods and pushed me back against the bumper and *vibrated my chest wall and pants legs* (emphasis added). I was stunned. Being a performing and recording rock-and-roll musician and a big game hunter of elk, bear, turkey, deer, and moose, and using calls to lure in these animals and hearing their calls in the woods, I had familiarity with both loud sounds and sounds of animals in the wild. This sound was no human, and no animal that I have hunted or know of."

Dr. Greg Bambenek (on left), clinical psychologist, who witnessed what may have been a sasquatch loud call. Also present: LeRoy Fish, Owen Caddy, and Rick Noll.

Dr. Bambenek's observation that the scream vibrated his chest wall and pants legs, raised a very interesting possibility that was also independently suggested by Benson. Could it be that sasquatch utilize infrasound in their vocal repertoire? The pitch of a sound is determined by its frequency, or cycles per second (Hz). Humans hear sounds within the sonic range between 20 and 20,000 Hz, although most adults' sensitivity lies between 40 and 15,000 Hz. The lowest note on a piano is 26 Hz. Infrasounds lie below the human range of perception. Humans cannot hear infrasonic vocalizations, but they can be felt if loud enough.

First recognized in whale communication, it was later demonstrated that land animals also employ infrasound in communications, and the list of species doing so continues to grow—elephants, hippos, rhinos, alligators, giraffe, lions, tigers, okapi, and several birds. For example, elephants have a hearing range believed to start at 0.1 Hz. Low-frequency sounds carry over considerable distances depending on their amplitude. They have low susceptibility to interference, which is a particular advantage to communication in uneven terrain and through dense vegetation. Infrasound communication functions best at night in clear dry air, probably due to strong refraction by inversion-like thermal gradients. Under ideal conditions an elephant's infrasonic call may carry over 30 km². This permits coordinated movement of dispersed family groups within the African savannah. Such ability would clearly be advantageous in maintaining social contact for a relatively solitary, wide-ranging primate in North American forests.

Infrasound has other adaptive uses in addition to communication. Recent research has shown that tigers stun their prey with an infrasonic roar prior to pouncing. The audible roar is mixed with infrasounds below 18 Hz, with the result that the roar can actually be felt. It has the effect of momentarily paralyzing the tiger's prey. This observation is quite interesting when considering a report of a sasquatch preying on a deer related by Dr. Bindernagel. "A man standing on the edge of a meadow at dawn saw a deer burst out of the timber into the open. A terrific scream sounded in the woods behind the deer. The deer froze, and huge bipedal creature appeared and grabbed the deer by the antlers. It picked it up and bit its neck, the deer offering no resistance. The creature stood for a moment looking about and growling. It then grabbed the deer by the muzzle with one hand, threw it over its shoulder and walked back into the woods."

Infrasound has been shown to produce a feeling of disorientation or a sense of uneasiness or fear. This may be the result of increased pressure in the middle ear caused by the infrasonic waves. Experiments have demonstrated that infrasound may affect the function of internal organs and disturb the sense of balance. Dr. Candace Pert, professor of physiology and biophysics at Georgetown University Medical Center, discoverer of the opiate receptor in the brain, has identified numerous neuropeptides that mediate among other things physiological and/or emotional responses. She recently announced that the receptors associated with these neuropeptides are responsive to specific resonant frequencies, many of which lie within the infrasonic range. This observation provides the physiological basis for the "visceral" response to exposure to infrasound. If sasquatch is capable of generating infrasound in response to human intrusion, it might account for John Mionczynski and another Fish and Game officer who were attempting to track a sasquatch and experienced a sense of disorientation and nausea to the extent that one of them vomited; or the numerous witnesses describing an inexplicable fear response, even to the extent that seasoned hunters refuse to return to the site of their encounter.

Infrasound can only be detected by special recording equipment and to date no surveys of possible infrasound use by North American wildlife have been conducted in the field. Could infrasound be a component of the vocal repertoire of sasquatch? It would certainly seem appropriate for their inferred distribution and would account for otherwise inexplicable aspects of their behavioral abilities.

Dr. John Mitani, a specialist in chimpanzee behavioral ecology at the University of Michigan, notes that many primates produce species-typical loud calls to communicate over long distances to attract mates or to maintain spacing within the group. Sound attenuates over distance due to atmospheric absorption, scattering, and interference due to waves reflected from the ground. Atmospheric absorption and scattering decrease with decreasing sound frequency. Low frequency sounds are less susceptible to interference. Dense vegetation further lessens the interference caused by ground reflection, as does the production of the sound from a higher vantage point. Semiterrestrial primate species typically vocalized from an elevated position. His research demonstrated acoustic adaptations to increase their propagation over distance. He found a decided trend for primates with larger home ranges to have loud calls with lower frequencies.

In general, primates generate sound energy most efficiently at frequencies near their characteristic resonant frequencies, which in turn vary inversely with the linear dimensions of the resonator. Therefore, the larger the primate, the lower will be the resonant frequency. To date no studies of the potential use of infrasound in great ape vocalization has been undertaken. However, when questioned on the matter, Mitani suspected that it was very probable that apes employed infrasonics in their calls. Given the large size of the sasquatch and its solitary and far-ranging habits, it might be expected that their loud call contains very low or even infrasonic elements.

Many primates have inflatable air sacs that extend out from the larynx. These take on distinct architecture in different groups of primates. Apes have a type of air sac that is referred to as lateral ventricular. They arise from the larynx just above the vocal cords and extend variably above the hyoid bone, along the jaw, over the breastbone and ribs and even to the armpit. The extent of the air sacs can be quite elaborate. The complex air sacs of the orang can hold up to 6 liters of air. The evolution and function of the air sacs have not been studied extensively. The best-supported hypothesis is that they serve as resonating chambers, acting to amplify vocalizations in the production of loud calls. It has also been suggested that they increase the duration of loud calls and alter the formant frequencies, enabling smaller primates to sound like much larger ones. It has been reported that air forced from the air sacs may set up vibrations in the ventricular folds that lie above the vocal cords. In this way apes could produce vocalizations during either inspiration or expiration, or even create overtones when combined with vibrations of the vocal folds or other regions of the larynx or pharynx.

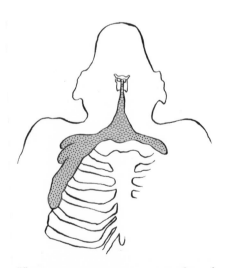

The great apes possess a complex of air sacs that extend from the larynx. (Reproduced from a figure by Daris Swindler and Charles Wood)

Chimpanzees have the least developed air sacs of the great apes, and humans have virtually lost them altogether. Occasionally there is a small saccule diverging from the ventricle of the human larynx that is the vestige of this anatomy. The implications, if any, of the loss of air sacs in the development of human speech are not understood at this time. However, there is an interesting human cultural parallel that may illuminate the function of the ape air sacs. A number of human cultures including ancient Egypt, Rome, and Turkey have in effect reinvented the air sac in the form of the bagpipe. Just as the ape air sacs provide a reservoir of air allowing sustained continuous vocalization, the bagpipe's bag serves a similar purpose. Of course, the highland clans of Scotland developed the instrument to its fullest and popularized the great highland bagpipe, or *piob mhor*. In the 1700s, a third drone, the great drone or bass drone, was added to the pipes. As a musical instrument of war, the Great Pipes were without equal. The shrill and penetrating notes, combined with the low frequency whirr of the bass drone, could be heard up to ten miles away, even through the din of battle. It has been said that the drones were tuned to create a dissonance that wreaked havoc with the horses of the English cavalry. It is very possible that the addition of the bass drones produced infrasonic frequencies with its peculiar effect on man and mount.

Another aspect of vocalizations attributed to sasquatch are their whistles. Recall that the *dsonoqua* is depicted with pursed lips, to reflect their peculiar whistling call. Among the great apes, the bonobos are adept whistlers. When nesting in the evening the air fills with their shrill soprano calls. Whistling has played an important role in human communication as well, whether in the form of toe whistles, hollowed out deer phalanges dating back more than 30,000 years ago, or in the complex whistling languages of many present-day populations. At least seventy distinct whistled languages survive in such places as Papua New Guinea, Mexico, Vietnam, Guyana, China, Nepal, Senegal, and a few mountainous pockets in southern Europe. Only a fraction of these have been studied

The great Highland bagpipes with the third bass drone can produce an unsettling tone that carries a great distance on the battlefield.

closely. The practice is found in cultures where individuals are often isolated from one another, and are most often used in mountainous regions or dense forests. Some whistles can travel up to six miles. These conditions seem to parallel those attributed to the behavior and habitat of sasquatch and suggest that whistling would be a very appropriate form of communication for a solitary primate inhabiting mountains covered in dense forest.

In 1972, in the High Sierras of northern California, witnesses recorded what are purported to be sasquatch vocalizations outside a remote primitive hunting camp. There had been encounters there before and since, but on this occasion exceptional recordings were made in the presence of a journalist from Sacramento. One intriguing aspect of the recordings occurs during an exchange of whistles between one of the witnesses and a sasquatch. The character of the sasquatch whistle was quite distinct from that of the human. The harmonics and formant of the whistle suggest that it was not made with the lips alone. In his analysis of the whistles Dr. Lynn Kirlin, professor of electrical engineering, then at the University of Wyoming, noted, "The formants and corresponding short vocal tract lengths found indicate the likelihood that the creatures could be able to whistle utilizing only a part of their vocal tract. If the creatures have a humanlike vocal tract, they might be able to whistle using the constriction between the two vocal cavities." In other words, the whistle came not from the lips alone, but likely was made by constricting the throat. The ability to sing from the throat is not common to humans and thus Kirlin concluded it unlikely that this was a case of hoaxing.

An extraordinary human parallel to this ability is demonstrated by the throat-singers, or overtone-singers. The most famous of these are the Throat Singers of Tuva, a region of southern Siberia, but it is also practiced in Tibet, Mongolia, and the Arctic of North America. Throat songs combine a low sustained fundamental pitch, like the drone of a bagpipe, with higher pitched notes that resonate above the drone. The singer employs alterations in the shape of the vocal tract to create the unusual tones and harmonics. Levine and Edgerton, in a study of the acoustics of overtone singing, discovered that in addition to the vocal folds, other organs are drawn upon to generate the second sound. These include the false vocal cords, the arytenoids cartilage, the aryepiglottic fold, the epiglottic root, and the tongue.

Ron Morehead, one of the witnesses to the vocalizations in the High Sierras, remarked upon hearing a recorded sample of the throat singers' performance that some of the sounds were precisely like what he heard outside their camp. He compared it to the sound of a huge tuning fork. Given this potential ability in trained human singers it certainly seems reasonable that such ability is plausible among the vocal repertoire of an upright sasquatch.

That there are cries in the night that defy identification by experts is a fact. Which, if any, of these vocalizations is attributable to sasquatch relies on the testimony of witnesses and their recordings. However, the suggestion that these vocalizations may incorporate qualities such as the use of infrasound and overtonation are intriguing questions that deserve further attention. These possibilities may explain some of the curious reactions by witnesses to sasquatch encounters. When considered in context, these vocalizations seem well suited to the adaptive behavior of a giant solitary North American ape, by facilitating interspecific vocal contacts over considerable distances in rough terrain, and by providing a mechanism for stunning prey and deterring human intrusion.

11

GRIN AND BEAR IT: MISIDENTIFICATIONS

When evaluating eyewitness reports, not only is the credibility of the witness on trial, but also their individual powers of observation and interpretation. Someone inexperienced in the outdoors may be quite convinced that any bump in the night or blur of dark fur in the twilight was nothing else than the legendary sasquatch. What is the likelihood that encounters in the wilds of North America ascribed to sasquatch could be simply explained away as cases of mistaken identity? The most likely candidate for potential confusion in the northwestern woods of the United States is limited to the American black bear (*Ursus americanus*), although in some regions of the United States and Canada grizzlies (*Ursus arctos*) are an additional possibility. The superficial resemblance of a bear standing on its hind legs to reports of an upright sasquatch are obvious. The musculoskeletal anatomy of the bear is remarkably humanlike. Many a hunter has been struck by the resemblance of a skinned bear carcass to a man.

The five-toed hind paws of a bear also display an uncanny similarity to human feet, save the presence of claws and the fact that the shortest toe of the bear paw is the inside or medial toe, opposite the condition of the human foot. Occasionally, skeletal remains found in the forest have been

Bears standing upright (Courtesy of Lynn Rogers)

initially mistaken for the remains of a missing person or long lost hunter or hiker, or for the elusive physical evidence of sasquatch, only to eventually be identified as those of a bear. One leading forensic anthropology textbook includes a discussion of the similarity of the skeleton of the human hand and foot to that of bear paws. However, even that scholarly reference confused the description by misidentifying the skeleton of a right fore and hind paw of the bear as being from the left side.

Recently, a photograph of a five-toed dismembered foot created a bit of excitement. The partially decomposed foot was found along a road in Oregon. Only a fringe of black hair remained on the skin and there were no visible claws. I determined it to be a bear appendage on the basis of the shape of the ankle and heel bones, and the arrangement of the toes. More discussion of these distinctions of the bear foot will follow later in the chapter.

In behavior and appearance, no other animal is more subject to anthropomorphism than is the bear. It apparently held a position of distinction in prehistoric cultures. Many Native American and circumpolar cultures consider the bear either an ancestor or a brother. They point to the similarity of their organs, their appendages, their ability to stand upright, and their intelligence and power. The bear figures prominently in many native ceremonial initiations and rites of passage. This long-standing theme was successfully transformed into a recent Disney movie, *Brother Bear*.

Dr. Lynn Rogers, bear biologist at Minnesota's Wildlife Research Institute, has been referred to as the "man who walks with bears." Based on his intimate familiarity with these ursine denizens of the woodlands, he considers the chances of mistaking a bear sighting for a sasquatch possible, but unlikely for a knowledgeable observer. In fact, it should be considered that the very opposite may more likely be the case. The initial reaction of many eyewitnesses to a possible sasquatch sighting is to rationalize the experience by assuming that they have simply seen a bear. Their inclination is to account for their experience within an existing framework of familiarity—"It was brown and hairy and upright, so it *must* have been a standing bear." The fact is, bears do have the ability to briefly stand upright on their hind legs, but they rarely walk for more than a few steps in that posture before dropping again to all fours. Their gait is usually halting and awkward on relatively short hind limbs, and their forelimbs do not alternately swing with each step, but

Dr. Lynn Rogers, the "man who walks with bears" (Courtesy of Doug Hajicek / Whitewolf Entertainment, Inc.)

Field guide-style contrast of the appearance of an upright bear and a sasquatch (Courtesy of John Bindernagel)

instead are held out forward in front of the body. They also have characteristic physical features that distinguish them from the typical description of sasquatch. These "field marks" that aid in quick identification include: prominent rounded ears atop their heads, long snouts, sloping shoulders due to their lack of collarbones (clavicles), and short legs.

Dr. John Bindernagel likewise concludes that regular misidentifications are unlikely, especially in the case of sightings lasting more than one or two seconds. He notes that unfortunately, most field guides do not provide information that would help a tentative eyewitness differentiate and identify the sighting as a sasquatch. For greater contrast he had a sketch made in the style of a field guide to illustrate those field marks distinguishing a sasquatch from an upright bear. The most important difference, notes Bindernagel, is the prominent broad shoulders in front or rear view, contrasted with the narrow sloping shoulders of the bear. The second most obvious distinction is the flat face of the sasquatch in side view, compared to the prominent projecting snout of the bear. Therefore, if a tall hair-covered figure, with broad square shoulders, flat face, and long legs, is seen walking upright across a road, it is very unlikely that a bear is responsible. Referring to the field guide–style sketch, Dr. Bindernagel notes, "With such an illustration before us, otherwise inexplicable sightings begin to make sense."

Even many professional wildlife biologists are often unprepared to accommodate credible reports of a giant ape in their neck of the woods. "As a North American–educated wildlife biologist," said Bindernagel, "I understand that I and my colleagues have had little or no exposure to the biology of great apes. After all, we don't think we have them here, so why bother when there are moose, deer, elk, wolves, cougars, and other 'normal' wildlife species to study and manage? The result of this omission is that, to most of my colleagues, reports of an animal resembling an upright gorilla throwing stones, beating its chest, breaking branches, and vocalizing loudly is too bizarre to make sense. As a result, such reports are normally discounted and almost never filed. Had we been more exposed to lectures and the literature regarding the great apes of Africa and Asia—and cognizant of their anatomy and behavior—we might have been much more open to such reports. I think I am correct in attributing the tendency of wildlife professionals to categorize sasquatch reports as false to our ignorance of great ape biology."

What of inexplicable large elongated five-toed tracks? A review of a typical field guide to animal tracks and signs once again reveals the most likely candidate is the bear. While many fleet-footed animals walk and run on the tips of a reduced number of toes in a *digitigrade* fashion, such as the coyote or deer, the hindfoot of the bear resembles a human's pedal appendage in its flat-footed, or *plantigrade,* appearance and its retention of all five distinct toes that make contact with the ground. The toes of a bear are arranged essentially in an uneven arc across the end of the foot, while the human toes are more or less aligned in an oblique toe row. In the human toe row, the "big" toe, usually also the longest toe, lies on the inside of the foot; the shortest or little toe is the outer toe. The order of the bear toes according to size is reversed. The inside toe of the bear paw is more like a small nonopposable thumb, with only two phalanges, or toe bones, rather than three as in the remaining digits. The outside toe is longer and substantially larger. In the case of the mistaken roadside foot, once I had determined that the foot was from the right side, based on the relationship of the anklebone, or talus, to the heel bone, or calcaneus, it was obvious that the inside or medial toe was the shortest and therefore the foot was that of a bear. Bear toes are also equipped with claws, and in soft substrate, such as mud or snow, they nearly always leave obvious claw marks. Still the claws can be seasonally worn down from digging in the soil, to the point that they may leave little trace in a footprint, even in soft soil, as was the case in a set of very large and exceptionally clear black bear tracks I cast in northern California. The presence or absence of discernable claw marks cannot be relied

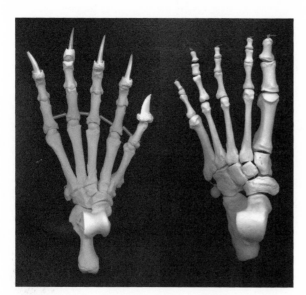

The foot skeleton of a bear (left) contrasted with that of a human (right). Both skeletons are from left feet.

upon as a distinguishing characteristic. In the case of a skinned or decomposed foot, the claws may be altogether missing.

Another distinguishing feature of the bear track is the shape of the heel. The heel pad of the bear is tapered to a blunt point and is usually separated from the interdigital pad by a prominent crease. This taper is so because the bear paw is not completely plantigrade, or flat-footed. Instead the heel is actually elevated slightly, lending the taper to the underlying pad. In large mature bears the heel pad may be a bit more rounded.

As in most quadrupeds, a greater portion of body weight is differentially carried over the bear's forelimbs so that the imprints of the forepaws are generally deeper. In contrast, the ape tends to carry more of its weight on its hind limbs and the heel bone is relatively thicker or more robust. With a true plantigrade foot, the ape or sasquatch foot has a broader rounder heel pad to cushion that point of weight-bearing beneath the calcaneus. Being bipedal, humans carry all of their weight on their hind limbs and walk with a striding gait and a characteristic heel strike at the commencement of the stance phase of a step. Therefore, they have relatively broader heel bones than a bear or an ape and thus an even more pronounced rounded heel outline. Given its extreme size and weight, the sasquatch has an even more pronounced heel and a broader and well-rounded heel pad than a human.

The pattern and spacing of individual footfalls also distinguishes human and reputed sasquatch tracks from bear tracks. Generally the four-legged or quadrupedal pattern of the bear track is evident in the series of alternating fore- and hind paws. At normal walking speeds, the paws are well separated and the interval between right

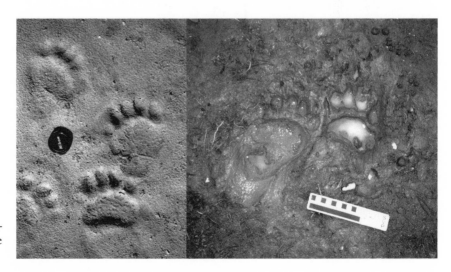

Tracks of black bears in northern California and Yellowstone National Park

Registered fore- and hind paws of a Montana grizzly bear (Courtesy of Roy Leland)

and left paws, or straddle, is pronounced. The bear's forepaw lacks a heel pad and merely has a wedge-shaped to bean-shaped interdigital pad. It is noticeably concave on its proximal border, giving it a bean-shaped outline. However, with age this concavity may tend to fill out somewhat. At swifter speeds the hind foot imprint often oversteps, overlaps, or is directly superimposed upon the forepaw imprint. The latter two circumstances give the impression of a single elongated footprint. This precise overlap or register of the fore- and hind paws happens infrequently and inconsistently, but can give the impression of tracks left by an animal walking on only two legs, although with a very short step length and noticeable straddle.

One such example of this was submitted initially as a possible sasquatch footprint by Roy Leland. Found in 1974 along the Yaak River in Montana, the combined track measured nearly 14 inches in length and appeared devoid of claws. A photo of the clearest print was published by a prominent researcher at the time with indications that the track could be an authentic sasquatch track. However, after further consideration and consultation with game wardens, Mr. Leland very correctly concluded that the track was that of a large grizzly bear. The pad of the forepaw is quite distinct, and the imprint of the outside toe of the forepaw remains clearly visible, while the remaining toes are obliterated by the tapered heel of the overstepping hind paw.

One clue to distinguish such an overlapping track is to look at the shape of the heel. If the overlapping forepaw occupies the position of the "heel" of the elongated footprint as in this case, the bean-shaped pad of the forepaw will usually exhibit a concave trailing edge instead of a rounded convex outline of a sasquatch heel. If the hind paw dominates the hinder part of the print, the characteristic tapering heel of the bear hind paw will be evident.

In spite of the clear distinctions, occasionally bear tracks have been attributed to sasquatch, generally by persons unfamiliar with wildlife sign, but even occasionally by experienced hunters or outdoorsmen. Some mistakenly rely on a single diagnostic characteristic, such as the presence or absence of claws to establish identification.

They presume that since bears have claws, a track that obviously fails to show claws cannot be a bear and therefore must be a sasquatch. Instead, the entire suite of characters should be taken into consideration when identifying tracks.

As another case in point, several young men were hiking in the Sierra Nevada mountains and came upon some faint tracks in the gravelly hillside. The pattern of tracks was suggestive of a biped. The possibility of discovering sasquatch tracks excited them, but maintaining some presence of mind, they took appropriate actions. They documented the event on videotape, marked the tracks and photographed them, collected a sample of scat along the track, and when they found a footprint with some distinct toe impressions and no obvious claws, they went for some cement and returned and cast it. All these materials were then sent to me at my lab for inspection. Upon examining the cast, the track appeared about 13 inches long and made in pebbly soil, which obscured some details. The photos of the flagged trackway indeed seemed to indicate a bipedal trackway, but with a rather short step. Some additional warning signs readily surfaced. What appeared to be the "big toe" wasn't very big or differentiated from the remaining toes. The toes were disposed in an arc across the foot. The "heel" was deeply impressed, but didn't seem to be connected to the forefoot. A closer look revealed a concavity on the backside of the "heel" and the dissociated forepart of the foot tapered to a point and showed a faint split between the interdigital pad and the true heel pad. It was a case of an overstep combination of the fore- and hind paws of a black bear.

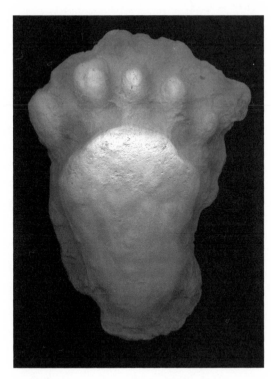

Cast of a bear hind paw associated with a sighting of "sasquatch" in southeastern Idaho

On the other hand, some authorities have been too quick to suggest that *all* supposed sasquatch tracks can simply be attributed to bears. Bruce Marcot, a research wildlife ecologist in Portland, Oregon, has maintained a Web page illustrating this line of reasoning, although stopping short of being dogmatic on the point. "It is my interpretation," he states, "that some, maybe many, tracks of Bigfoot and its global cousins, as reported earnestly by honest people (thus, excluding the hoaxers), are probably tracks of large wildlife—and most likely bears that are simply enlarging their own tracks by sliding, by overstepping, or by having their tracks enlarged through weather effects." After superimposing the unscaled outline of an overlapping brown bear track onto a photo of an alleged sasquatch track cast he concludes, "In general, the shapes seem to fit rather well, allowing for some minor realignment of the toes. The size of brown bear tracks is larger than those of American black bears, so this little experiment should be redone with more precise measurements of actual casts and tracks. But the shape here seems to fit rather strikingly." I have shown that indeed one needs to be wary of this hazard for misidentification. In this case, however, the "fit" described by Marcot is actually quite superficial and quickly falls apart when the aforementioned details such as absolute size, toe proportions, heel shape, claw marks, pads, and so on, are carefully examined. Furthermore, a number of the examples depicted by Marcot employ my own photos of alleged sasquatch footprints from the Blue Mountains, which I examined firsthand within a long line of clear bipedal tracks, as discussed in the Introduction. In this instance there is no question of misidentified bear tracks. Thus, whatever merits remain in Marcot's cautionary note, there is obviously danger in overgeneralizing from an isolated image of a footprint, particularly when specific details are glossed over.

Dr. Rogers is of course very familiar with bear tracks as a result of his many years of experience studying bears. He has also personally examined casts of alleged sasquatch footprints. As he hefted a large, exceptionally clear cast from the west coast of Washington that was collected by an on-duty deputy sheriff, he observed, "A lot of times bear tracks are reported as people tracks [and sometimes sasquatch tracks], but I cannot explain this track. I don't know how they would fake it. The big toe compared to the other toes in a humanlike pattern is not like a bear at all. I just cannot explain this."

There is no question that footprints and upright hairy figures have been and will be occasionally misidentified, but this does not explain the numerous credible sightings by knowledgeable observers and the persistent examples of distinct footprints. It should be obvious to any experienced tracker or expert in foot morphology that not all footprints attributed to sasquatch can be casually dismissed as bear footprints, nor can the observations of experienced outdoorsmen and biologists be brushed aside as simply cases of misidentification.

12

By the Numbers: Statistical Analyses

As a young newspaperman, John Green was aware of stories of the hairy giants of Canada's Pacific Coast. First popularized in 1929 by writer J.W. Burns, a teacher on the Chehalis Indian Reservation near Green's hometown of Harrison Hot Springs, sasquatch had become something of a regional icon and the focus of the town's celebration of Canada's Centennial. Never having taken the "Indian legend" seriously, Green was nonetheless intrigued by firsthand accounts of giant footprints by people he knew well and respected. Said Green, "Interviewing people and gathering facts are my regular occupation, and if I were being fooled very often, my readers would be bringing it to my attention. These stories rang true to me, but I took the additional step of having them legally sworn to . . . I took what I had gathered to the University of B.C., expecting that scientists would want to investigate the matter. I still had a great deal to learn." Undeterred by the offhand dismissal of the possibility of a giant ape, he went on during the succeeding decades to collect accounts, numbering into the thousands, of encounters with hairy giants and discoveries of huge footprints by people from every walk of life.

What is the utility of Green's collection

Young newspaperman John Green shows a 17-inch cast of California's Bigfoot to residents of Agassiz, British Columbia. (Photograph by Jack Long)

of stories? Taken individually we are not only at the mercy of the veracity of the story-teller, but also limited by the witnesses' individual power of observation and accuracy of interpretation. Furthermore, the compiled data do not represent a systematic, regimented sampling taken under controlled conditions. These data represent the sum of scores of serendipitous encounters, which may or may not be credible, that happened to get reported and happened to make their way into the files of John Green. Many of these came by way of newspaper clipping services or radio news reports, personal correspondence, and published narratives. Personally investigating and assessing the credibility of each and every report would fully occupy the attention of a large full-time investigative team with an unlimited budget, let alone a single journalist with a job to hold down and a family to support.

However, upon reflection it will be recognized that this sort of anecdotal data forms the basis for many valid statistical analyses. Green notes, "If it is any comfort, I once did a study of a wide variety of points, comparing percentages obtained while using only those stories which I thought should be reliable, and those obtained using every report in the file. The differences were insignificant in almost every case." Therefore, there appears to be an internal consistency to the signal emerging from the data, but whether this is a signal based on a common perception of a real animal or an imagined one is the question. Might there possibly be an unconscious filtering of accounts that fail to conform to some preconception of an American abominable snowman?

What is the hypothetical portrait of the sasquatch and its behavior that is implied by these reports? In Green's own words, "The reports are numerous enough to establish a few things about the sasquatch lifestyle. They are omnivorous, with almost equal mention of meats and vegetable matter in observations of things eaten, or apparently taken presumably to be eaten. They are largely nocturnal. Although humans cannot see well in the dark and there are far more humans around in the daytime, almost half of the sightings take place at night. They are not active in cold weather. Less than 10 percent of reports mention snow, and tracks in snow are rare. They have an affinity for water. Unlike the known apes, they have been reported swimming, both on the surface and underwater.

"The reports are also informative in what they do *not* mention. In spite of a common assumption that sasquatch live in caves, indications of use of caves, or any form of shelter, are very rare. Tool use is not indicated at all, and while objects are sometimes thrown, it is in a looping underhand manner, not in an overhand straight line. There are also no reports of fangs, or claws, an unlikely omission if we are dealing with an imaginary monster. These are presumably not reported because they don't

Sasquatch tracks in snow in the foothills of the Blue Mountains near Walla Walla, Washington. The compacted snow beneath the footprints has pedestalled as the wind has blown away surrounding snow. Their size and length of step is contrasted to the smaller and parallel human track. (Courtesy of Brian Smith)

exist, but there are also very few reports of females, infants, or small juveniles, which must exist. This brings into question one of the most obvious assumptions, that sasquatch are solitary animals. Less than 10 percent of reports involve more than one creature, but if females and their young are very rarely seen it remains possible that family groups exist, while normally only lone males take a chance of encountering humans."

Reported sizes of sasquatch encountered range from under 4 feet to nearly 14 feet, with an average height of 7.5 feet. Hair was straight and short to medium length. Half of the reported colors were black or dark, one quarter were brown, 15 percent were light in color and 8 percent were gray. Only 14 percent of the reports made any mention of a noticeable smell. Posture was usually erect or with a stoop.

Is it possible to distinguish natural patterns in such observational data concerning sasquatch, from the inherent randomness that one might expect of fabrications and hoaxes? Are there any patterns within the data that would be suggestive of a living population rather than a fictitious phenomenon? Dr. Henner Fahrenbach, retired microscopist, formerly with the Oregon Regional Primate Research Center, has published a statistical analysis of reported sasquatch dimensions based in large measure on Green's compiled data. He limited his analysis to those reports emanating from the western U.S. and Canada. Dr. Fahrenbach states, "A sufficiently large sample size, as we have for many measurements, has the advantage of supporting through such treatment either an origin from a living population of animals or, conversely, exposing a set of fictitiously generated values." With a large enough sample, it can be presumed

Dr. Henner Fahrenbach was among the first to publish a statistical analysis of metrics of sasquatch anatomy. (Courtesy of Doug Hajicek / Whitewolf Entertainment, Inc.)

that any biological patterns present may emerge from the extraneous outliers.

First off, Fahrenbach examined patterns in the frequencies of footprint dimensions. He noted that footprints are the standard stock in trade of sasquatch research, and have the advantage of providing a degree of objectivity not always attainable with other categories of observational data. Even so, evaluating footprint dimensions is a different matter than say, taking standard measurements of bones in a museum cabinet. An individual footprint is the trace evidence of the interaction of the foot with the variable conditions of the substrate. Apparent dimensions can be influenced and distorted by the dynamics of walking, e.g., slide-ins, drag-outs, curled toes, depth of impression, foot rotation, etc. The measured greatest length of individual footprints even in a single trackway may vary by as much as 10–15 percent. All of these factors introduce intra-individual variation that will affect derived correlations and indices.

A large proportion of continuous frequency distributions in nature approximate a normal distribution, a symmetrical bell-shaped curve. It would be expected that if reported footprint lengths were indeed sampling a living population, the summation of that data would also approximate a normal dustribution. Fahrenbach found that a sample of sasquatch track lengths distribute in a somewhat peaked (kurtosis = 1.89) bell-shaped curve. The range of variation in footprint lengths included from 4 to 27 inches, with a mean of 15.6 inches long and a standard deviation of 3.1 inches. The distribution does not suggest extreme sexual dimorphism, or marked difference in size between males and females, which would be reflected in a bimodal distribution. Fahrenbach acknowledges, "At most, it can be speculated that, at the median of footprint size, the sexual dimorphism does

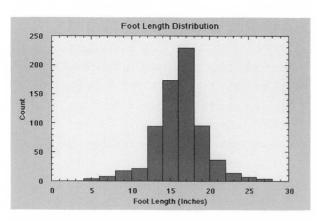

The distribution of foot lengths derived from a sample of reports (Courtesy of Henner Fahrenbach)

Examples of the smallest footprint casts in the author's research collection

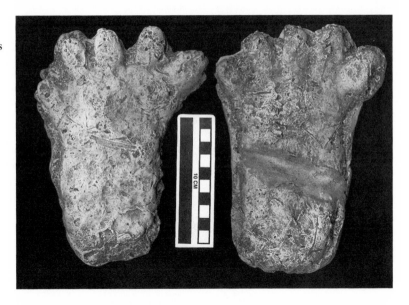

not exceed two inches in foot length or roughly one foot in body height" (since the measurement data were summarized in 2-inch intervals). It should be recognized that the sample is not necessarily limited to fully adult individuals. A slight negative skewness of the distribution suggests the contribution by juveniles presumably included within the sample. Fahrenbach reasons, "The normal distribution overall argues compellingly against any alternative hypothesis to the existence of the sasquatch as a cryptic species, in that production of fictitious data over forty years by hundreds of people independently of each other would have generated a distribution with many peaks."

Fahrenbach continues by adding related dimensions of the foot and gait that are correlated in a functionally cohesive fashion. For example, sasquatch footprints are not simply enlarged facsimiles of human footprints. They are consistently relatively wider than human footprints of such a size would be. The ratio of width to length is called the width index. Almost the entire set of sasquatch footprints for which length and width measures are available, has a greater width index than average human feet do. This relationship is consistent with the heavy body build reported for sasquatch. When the width index is scaled against foot length, a negative regression is revealed for combined male and female samples of both human and sasquatch feet. That is, the foot width increases at a slower rate than foot length, in both populations. In humans there is a more dramatic drop off in relative foot width, that is, people with very long feet typically have relatively narrow foot widths. In the sasquatch sample the index remains closer to 0.5 throughout the range of foot lengths, that is, the foot maintains a width that is about half its length throughout the size range.

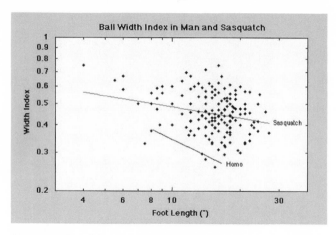

Bivariate plot of foot length to width for a sample of sasquatch footprints and human feet (Courtesy of Henner Fahrenbach)

Owen Caddy has independently confirmed this observation for a smaller but more recent set of data. A sample of forty widely varying human feet was measured, including foot length, width of ball, and width of heel. A subset was then impressed in a wide variety of soils to varying depths while in a variety of motions. Casts were made and similarly measured. Variances between measures of the foot compared with measures of the casts were less than 3 percent for foot length and less than 2 percent for measures of width. Next a sample of twelve putative sasquatch casts was measured. By expressing the shape of the foot as unitless indices and presenting the data as a bivariate plot, Caddy has demonstrated that there is virtually no overlap between the putative sasquatch footprints and the sample of human feet, even allowing for a margin of error of 3 percent.

It is clear that there is a slightly larger range of variation in the sasquatch sample. On this point Fahrenbach notes, "It is very probable that the sasquatch population, being composed of animals with a long life span, few offspring, and little culling by predators, will display wide physical variation."

In exploring the correlation of height and foot length, Fahrenbach began with the Patterson-Gimlin film subject. Patterson himself estimated the subject to be 7' 4" (other

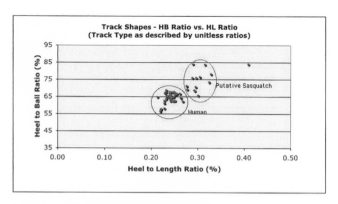

Bivariate plot of heel breadth ratios for a sample of sasquatch and human feet (Courtesy of Owen Caddy)

estimates based on the scale established by the 14.5" footprints have been lower). Jeff Glickman, a forensic analyst, arrived at a height of 7' 3.5" using a superimposition method with a graduated pole as a reference. The combination of a 14.5" foot length and 7' 3.5" height yields a foot length-to-height ratio of 16.5 percent, or a foot-to-height multiplier of 6.04. In addition Fahrenbach's data include eighty-nine reports with both measured foot length and estimated height. The resulting correlation (correlation coefficient = 0.558) yielded a slightly lower mean multiplier of 5.84. This relationship between foot length and height illustrates that the sasquatch foot grows disproportionately in excess of general bodily dimensions. Since body weight, a function of volume, increases to the cube of linear dimensions, and surface area of the sole of the foot increases to the square, it is to be expected that foot length (and breadth) scales positively with increasing height. In other words, the foot gets relatively longer at a greater rate than height increases. The allometric formula describing this disproportionate size relationship yields a height estimate of 7' 7" for a 14.5" foot. This is slightly higher than the estimate for the Patterson-Gimlin film subject, but considering that substantial individual variation is anticipated, and that the sample presumably pools male and female (as well as juvenile) individuals, the correlation is impressive.

Frequently, footprints are impressed to remarkable depths, raising questions about the weight of the sasquatch. Fahrenbach notes the difficulty of visual estimates of weight and instead proposes an estimator that utilizes the allometric relationship between chest circumference and body weight for primates ranging in size from tamarins to baboons, to which he added a sample of ten gorillas. The modified formula incorporating gorillas has a correlation coefficient of 0.956. Turning again to the Patterson-Gimlin film subject, Fahrenbach derives an estimate of chest circumference of 60". When the scaling formula is solved using this estimate, a predicted weight of 542 lbs is the result. Considering the variation within the gorilla sample, the potential range of values for the Patterson-Gimlin film subject lies between 400–700 lbs. Extrapolating further, it could be predicted that a sasquatch with the mean foot length of 15.6" would stand 7' 10" and weigh 658 lbs; an 18" footprint equates to a 8' 4" sasquatch weighing 768 lbs. Bear in mind that there is considerable variation around these predicted values to be expected in a natural population.

The correlation of foot dimensions with predicted weights has significant implications for the accommodation of plantar pressures beneath the foot of a giant bipedal ape. Bone, cartilage, and fatty connective tissues of the foot can only tolerate limited imposed pressures. Allometric relationships, or disproportionate scaling of body dimensions, are adaptations to these stresses imposed on the feet by increased body weight. Fahrenbach examined several sasquatch footprint casts of various sizes, matched them to predicted body weight, and then calculated sole pressure values that varied from between 6 to 9 lb/in². By comparison, the human foot yields values of about 10 lb/in² for the contact area of the foot (the typical one-sided hourglass outline of an arched foot on a hard surface) and 5 lb/in² for the entire area of the sole of the foot. A booted foot has a value of only 2.5 lb/in². Thus, the estimated plantar pressure values for the sasquatch foot are not too different from those values for human feet (although 4–5 times that of a booted foot in the field). Excesses in plantar pressure are likely compensated for by behavioral adaptations and modifications to the soft tissues of the foot, for example, a more compliant gait, flat feet, and thicker sole pad. Recent studies of human locomotion have revealed that a compliant limb posture alone may reduce stresses by as much as 18 percent. A relatively broader, thicker heel pad has considerably increased fat volume to accommodate increased loads on the foot.

Could random independent hoaxes produce such a well-correlated pattern? Fahrenbach concluded, "I have come to the following conclusion. The histogram that resulted from, for example, the foot length, is indicative of originating from a living population by virtue of its bell shape. If it had been fictitious, you would never have gotten this sort of shape, which is distinctly peaked at 16 inches. Therefore, I must conclude that in all probability there is a living entity that has produced these data—a large primate walking about in the North American forests . . . Study of this rather overwhelming mass of circumstantial evidence would suggest that scientists should take this matter more seriously and not reject it offhand."

The geographical pattern of distribution of reported encounters with sasquatch or the discovery of footprints holds potential for additional analysis. Recent computer technologies have permitted the development of GIS (geographic information systems), which is used to capture, display, and analyze spatial data information. GIS enables one to envision the geographic aspects of a body of data and receive the results in the form of some kind of map. These various datasets are referred to as coverages and may contain data on soil types, vegetation types, precipitation, wildlife ranges, etc.

There has been little attempt to apply this novel GIS approach to the analysis of sasquatch reports. An initial attempt was undertaken by Peter Aniello as a student at the University of Colorado at Denver. Aniello utilized data available through the Big-foot Field Researchers Organization (BFRO) and established coordinates for 377 encounters within the Cascade Bioregion in Washington, Oregon, and northern California. An objective of his analysis was to determine if any trends in the spatial distribution of these encounters would reflect movement of the sasquatch "population" correlated with human population expansion, especially in the regions of Seattle, Washington, and Portland, Oregon. This assumes that all of the sasquatch in the Pacific Northwest constitute a single population and behave as such, which Aniello acknowledges may or may not be the case. The patchiness of their distribution may contradict this assumption. It should also be recognized that the encounter data are not the result of a systematic survey or transect. They are the accumulated record of intersections between spheres of human and presumed sasquatch activity. Nevertheless, Aniello's analysis produced intriguing results that revealed trends suggesting that as a whole, sasquatch encounters were occurring farther from population centers and at higher elevations.

Aniello concludes, "GIS technology is a valuable tool in wildlife management of known animals. As this study demonstrates, it can also be a valuable tool in tracking movements and characteristics of unknown animals, or cryptozoids, possibly giving clues to these animals' whereabouts and habits. The ability to uncover and analyze hidden spatial patterns in data using GIS is perhaps its greatest strength, and should be an integral part of future sasquatch studies, as the subject gains legitimacy."

13

STEPPING THROUGH TIME: THE EVIDENCE OF FOOTPRINTS

My research interests revolve around the functional anatomy and evolutionary history of the primate foot as it relates to the otherwise singular human adaptation of walking on two feet, i.e., bipedalism. The skeletal and muscular anatomy of the foot, from primitive mammals and monkeys to great apes and humans, has preoccupied a large part of my professional career. So it was the evidence of the sasquatch footprints that particularly captured my attention. Clearly, *something* is leaving enormous humanlike tracks on the backcountry roads and riverbanks of North America's mountain forests. With the potential for misidentification of bear or human footprints accounted for, the otherwise inexplicable footprints that remain must be either hoaxed or hominoid. In the absence of bones or body, the tracks constitute the most abundant and informative data that can be dealt with by scientific evaluation. These traces left in the ground are for me a sizable chapter in the story behind the sasquatch, and one that is often underappreciated.

For hunter or conservationist, tracks left by animals leave clues about their numbers and behaviors. Renowned wildlife tracker Dr. Jim Halfpenny observed that mammals are among the most elusive animals in the world. Much of what is learned about mammals in the wild comes from the stories that can be read from their tracks and other sign. I spent an afternoon with Halfpenny in his Gardiner, Montana, home just north of Yellowstone National Park, discussing the interpretation and nature of sasquatch tracks. We jointly examined a small selection of sasquatch casts that I had brought along to illustrate the traces of dynamic features I observed—toe movement, midfoot flexibility, pressure ridges, etc. These duplicates were subsequently left with Halfpenny to be added to his remarkably extensive and informative collection of wildlife track casts. He was clearly impressed by the animation evident in the

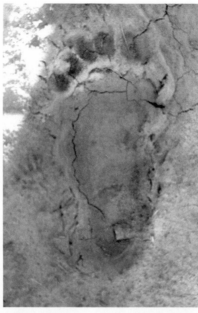

A clear example of the sasquatch footprints located by Deputy Sheriff Dennis Heryford in Grays Harbor County, Washington, in 1982 (Courtesy of Dennis Heryford)

sasquatch track casts, and interested in my inferences about their dynamics, but in the end he seemed to feel that the story told by the tracks did not yet have a definitive conclusion. He was quite willing to give serious consideration to any new track evidence we might recover, and challenged me to put him onto a fresh track. Then he was confident he would get to the bottom of it one way or the other.

Dr. John Bindernagel echoed many of Halfpenny's sentiments and further addressed the significance of sasquatch tracks: "There is so much more evidence for the existence of sasquatch than most people realize. In addition to eyewitness descriptions and drawings, we have hundreds of descriptions, photographs, and plaster casts of the tracks. For me as a wildlife biologist, it's the tracks that we depend upon for the existence of an animal in a study area. We don't usually see the mammals, but we do see their tracks. In the case of the sasquatch, this is the most compelling evidence we have." Clearly those who have most closely examined the footprints attributed to sasquatch are the least inclined to simply dismiss them offhandedly as hoaxes.

The sometimes-enormous size of the sasquatch tracks gave rise to the common American appellation of "Bigfoot." These footprints average between 15 and 16 inches in length, with a reported range of 4–27 inches. Their superficially humanlike appearance is largely the consequence of the inner big toe being aligned with the remaining toes, whereas an ape's inner toe diverges much like a thumb. The resemblance to human footprints largely stops there, however. In fact, the sasquatch footprints lack the principal distinctive features that set the human foot apart from that of its hominoid cousins. Sasquatch footprints are typically flat with no consistent indication of the true hallmark of the human foot—a fixed longitudinal arch. Additionally, there is little indication of differential weight bearing under a specialized "ball" at the base of the big toe. The sasquatch foot is relatively broader and the sole pad apparently thicker, by comparison to human feet. The heel and toe segments are disproportionately longer.

The tracks do indicate a bipedal gait, one where sasquatch walks on two feet, long considered a strictly human trait, but now recognized to be widespread among prehuman hominids, and perhaps present even in a much earlier Eurasian great ape called *Oreopithecus*. This latter case has prompted recognition that the evolution of bipedalism is not a singularly human trait, but a manner of moving on the ground that has evolved independently in parallel, multiple times. Indeed, apes display a clear tendency to adopt an upright posture and to walk on two feet from time to time— referred to as *facultative* bipedalism. It appears that more than one species of ape became a full-time biped, or *habitual* biped. This further blunts the criticism of the hypothesis that *Gigantopithecus* was perhaps bipedal, evolving the habit convergently under similar ecological conditions.

The manner of walking on two legs inferred for sasquatch appears quite distinct from the typical mode of modern human walking, which may reflect an independent origin. The succession of sasquatch footprints frequently displays an exceptionally long stride, often with the footprints marshaled one directly in front of the other. Although human footfall patterns exhibit a considerable range of variation,

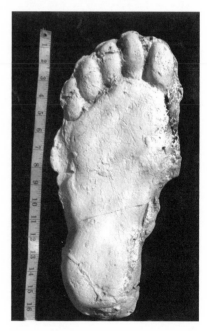

One example from a series of casts made by Officer Dennis Heryford

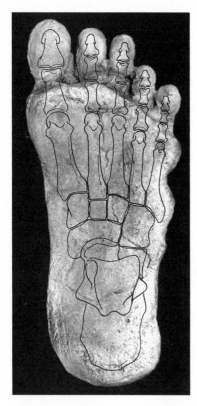

From the shape of the foot a reasonable inference of the skeletal proportions can be made as in this cast from Grays Harbor County, Washington.

most modern Euro-Americans leave lines of alternating right and left footprints separated by some distance referred to as a straddle, or step width. Sasquatch footprints generally lack the indications of differential pressure beneath heel and ball as in an arched foot, but instead are rather uniform in depth. Occasionally, the dynamic signature of the foot indicates a more apelike midfoot flexibility. In all, the sasquatch footprint is not merely an enlarged facsimile of a human footprint, but appears to represent a uniquely adapted primate foot associated with a distinctive mode of bipedalism, one that may well have evolved independently although roughly in parallel to hominid bipedalism. What is this exceptional sasquatch foot suited for?

As an anatomist and physical anthropologist at Idaho State University, I have examined fossilized hominid footprints from sites as diverse as Hawaii, Turkey, France, and East Africa. Turning my attention to the sasquatch footprints, I now have assembled nearly two hundred casts of sasquatch tracks in my lab, including the addition of the late Dr. Grover Krantz's cast collection. This assemblage has provided a significant opportunity to assess the degree of consistency as well as the variability present among these footprints. It has also highlighted the occassional hoaxes and misidentifications.

Dr. John Bindernagel was an invited speaker on the subject of sasquatch at the campus of Idaho State University in 2002, and on that occasion he was able to inspect and study the assembled casts housed in my lab. In a subsequent letter to the dean of Arts and Sciences he shared the following sentiment: "The opportunity to spend three days in Jeff Meldrum's lab during this period gave me an opportunity to examine the quality and extent of the material he has assembled. With his permission, I will be directing other scientists to this collection. I can only commend Jeff Meldrum and ISU for addressing this subject in a timely and appropriate manner, showing scientific leadership, which has been sadly lacking up to now in this subject area. Based on the results of my own research I am confident that such leadership will eventually—and perhaps sooner rather than later—be amply repaid as other scientists finally accept our responsibility to address this subject. The ISU collection of track casts requires biologists to explain it, and must figure importantly in upcoming discussions."

As I embarked on a review of this sample of footprints, a number of questions came to mind. Can the repeated appearance of recognizable individuals be demonstrated by the presence of their footprints in a particular region? Are there documented series of footprints that exhibit animation through variation in toe position and/or foot rotation as the living foot supports body weight during a step? Do the footprints exhibit signs of the dynamic interaction of the foot with the soil, such as pressure ridges and tension cracks? Are the footprints simply enlarged human footprints, or are distinctions present in the inferred anatomy that make sense considering the sasquatch's large size and habitat preferences? What can be learned about any underlying consistency and variation from this truly unique sample of footprints?

My first consideration was the repeated appearance of particular individuals. It stands to reason that if these animals are as rare as they must certainly be to have remained so elusive, then when footprints *are* found in a given region over a period of time, chances

Roger Patterson displaying two casts of sasquatch tracks (Courtesy of Vance Orchard)

are they were left by one of a very few local individuals and there should be repeated appearances of recognizable footprints. After all, the five-toed soft-soled footprint of sasquatch would reasonably exhibit more distinguishing individual characteristics than a hoofed-deer track. This has certainly been borne out by assembling as many

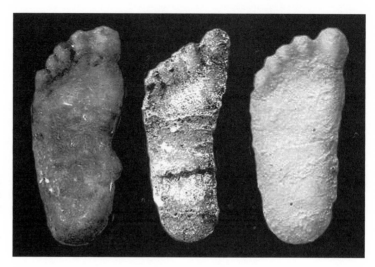

Repeat appearance of a particular individual sasquatch as attested by this series of casts made at the Bluff Creek construction site by Jerry Crew, 1958; Hyampom, by Bob Titmus, 1962; Bluff Creek above Notice Creek, by Roger Patterson, 1964

specimens of footprints as possible with information about the circumstances of their discovery. A case in point would be the original "Bigfoot" cast made by Jerry Crew at Bluff Creek, California, in 1958. Tracks of that individual with a 16–17-inch footprint were seen a number of times over the subsequent years, including a series of casts made by Bob Titmus and Syl McCoy near Hyampom in 1963, and a pair of tracks cast in the region by Roger Patterson in 1964 at Laird Meadow above Notice Creek, not far from the original Bluff Creek site.

A number of other examples soon became evident as well, including a notable similarity between the 15-inch track from the Blue Creek Mountain Road in 1967 and the casts from the Patterson-Gimlin film site. Investigators at the time felt that the casts Roger made were different from the 15-inch track that had been seen on a number of occasions. This impression may have resulted from the difference in nature of the substrate that the tracks had been observed in, ranging from firm moist sand to powdery dust on a hardpan logging road. A careful comparison reveals a high degree of congruence. Carefully outlining and superimposing scaled images of the prints and casts can readily visualize this resemblance. The implication is that the subject of the Patterson-Gimlin film also left tracks on the Blue Creek Mountain Road a month earlier. In fact this individual appears to have left footprints in the region on a number of occasions.

Another example includes several casts and photographs of footprints that appeared to be from the same individual that left the tracks I observed in 1996 in the foothills of the Blue Mountains in southeastern Washington. In Krantz's collection was

On the left is a footprint from the Blue Creek Mountain Road trackway in 1967; in the center is a cast made by Roger Patterson at the film site just over a month later; on the right are the superimposed scaled outlines of the footprint and cast, suggesting they were made by the same individual.

a copy of a cast made in northern Oregon, about twenty miles south of the location of my footprint site. Two local anglers happened upon a string of five 14.5-inch footprints along the stream bank in a densely timbered area seldom frequented. One footprint was impressed an inch deep in mud and was quite distinct. The others were in grass and retained less detail. A matted down area in the ferns indicated a large animal had lain there and a nearby 3.5-inch yew was snapped off four feet above the ground. They mentioned their find to the proprietor of a local store. Although incredulous, she and her husband went out to have a look for themselves. Their dog, a dingo-shepherd cross, acted terrified by something in the area. They photographed the print, covered it, and returned later to make a cast. The original is on

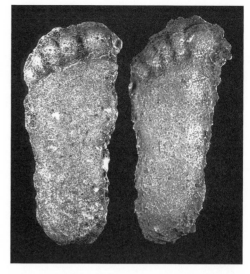

15-inch casts by Bob Titmus from Bluff Creek in 1958 (left) and the Patterson-Gimlin film site in 1967 (right)

display in their establishment and I have had the opportunity to examine it carefully and later add a duplicate to my collection. It corresponds very closely in size and proportions to the casts I made near Walla Walla, down to the unusual feature of the long second and third toes. It also displays midfoot flexibility about a transverse axis.

As an interesting aside, the Forest Service was contacted by phone and informed of the tracks. They simply laughed and asked what the witnesses had been drinking. They declined to investigate. However, a few minutes later one of the personnel from the office called the witnesses back and quietly asked if he might be shown the tracks and be permitted to photograph them. It turned out that some interested individuals in the office maintained an unofficial file of reports of tracks and sightings that occurred rather regularly in the region.

Ultimately, one could argue that repeated appearances might more simply be explained as repeated hoaxes by a persistent prankster, who periodically

Footprint and cast made near Tollgate, Oregon, 1986 (Courtesy of Lorretta Nietch)

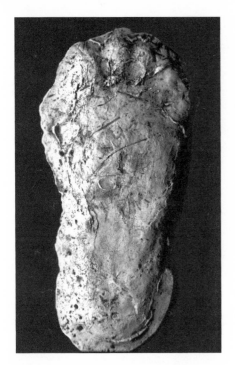

14-inch footprint cast by the author at Five Points, outside of Walla Walla, Washington

trots out a single pair of fake feet and lays down some bogus tracks, even if this involves widely separated and out of the way locations. This simplistic suggestion has some merit since there certainly have been instances where hoaxes have been exposed. These relatively rare instances are usually transparent, which brings us to the next factor—animation. A line of footprints made with artificial feet appears quite unnaturally monotonous to the experienced eye. There is minimal variation from one footprint to the next. Toe position is unvaried, mud cannot squeeze between joined carved toes. The intrinsic alignment of foot segments is constant, devoid of the signs of flexibile linkage present in a living foot, allowing it to dynamically interact with the ground under varying conditions of the soil and speed of walking.

The primate foot skeleton is composed of twenty-six individual bones, with nearly as many joints. These joints, especially those of the tarsus, the bones at the ankle end of the foot, have multiple axes of rotation defining their interrelated movements. This complexity permits the foot to conform to the inclined and uneven surface of the ground as the animal transmits weight through a step and moves over its environment. However, the skeleton is a linked system, and the shape of joints and the presence of ligaments functionally limit the potential movements of the elements of the foot relative to one another. In other words, there is a limited range of coordinated configurations that can be adopted by the living foot skeleton. For example, the axis of the foot may exhibit different extents of rotation, i.e., pronation or supination, which raises or lowers the inner edge of the foot as it twists about its long axis. Other signs of animation include variation in toe position, with differential degrees of flexion/extension and splaying. Such features of animation provide important clues about the form and function of the foot.

It is crucial to realize that the footprint is not simply a static mold of the foot, but rather a record of the transient dynamic interaction of the foot with the ground. In addition to the impressed shape of the individual parts of the foot, the forceful transmission of weight throughout the course of the step may produce characteristic distortions of the surrounding soil, such as tension cracks or pressure ridges that are quite distinct from the impact features of a weighted prosthetic or artificial foot. Likewise, distortions

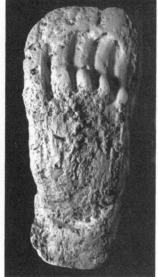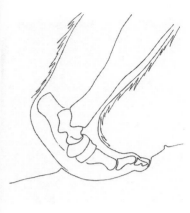

A half-track imprint and cast by the author at Five Points, outside of Walla Walla, Washington, and interpretation of the foot action that produced it

of elements of the foot can result from slide-ins or drag-outs, varying depth of impression, or by the expansion of the soft-tissue sole pad under weight.

Returning to the clear footprints I observed in a muddy road in the Blue Mountains of southeastern Washington, in addition to the evident animation, one of the most significant aspects was the indication of pronounced flexibility in the midfoot. In several instances, the heel had left no impression in the ground, presumably elevated in the air as when a sprinter runs on the ball of the foot. However, lacking a fixed arch, the sasquatch foot would flex much as an ape's foot flexes at the midfoot, specifically, at the midtarsal joint. Thus, only the forefoot, distal to the midtarsal joint would leave an imprint.

I commenced looking for additional examples of this expression of midfoot flexibility, and a number of further instances of similar "half-tracks" have been subsequently identified. One

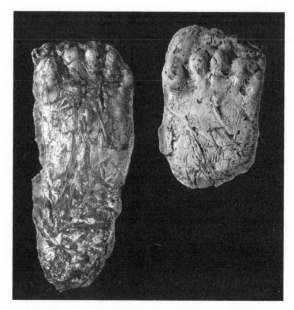

A 14-inch footprint cast by the author alongside a half-track

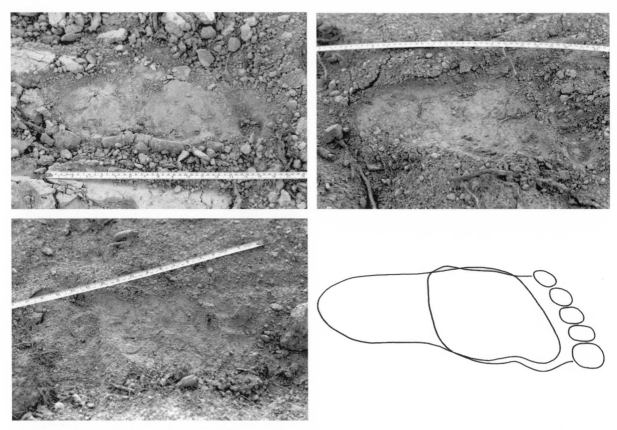

Series of 15-inch footprints examined on the Blue Creek Mountain Road, northern California. The upper pair are complete footprints, while the lower is a half-track. Notice the point of midfoot flexion revealed by superimposing the outlines of the prints. (Courtesy of John Green)

of these was documented by a set of photographs taken by Don Abbott, an archaeologist from the British Columbia Royal Museum, in 1967. These 15-inch footprints were part of an extensive trackway comprising over a thousand footprints along the Blue Creek Mountain Road in northern California and were examined thoroughly by John Green and René Dahinden, sasquatch investigators also from Canada. There were several photos of clear complete footprints, but one photo showed an abbreviated footprint, missing a heel imprint. The photos included a tape measure for scale and each was taken from approximately the same distance and angle. It was a simple matter to project, trace, and superimpose the outlines of the footprints to demonstrate that the relative position of the proximal edge of the abbreviated footprint appeared very similar to the half-tracks I had examined in the Blue Mountains of southeastern Washington.

Another striking example came from the west coast of Washington, in the eastern part of Grays Harbor County. Deputy Sheriff Dennis Heryford was one of several officers investigating footprints discovered near a spot called "Abbot Hill" in 1982. Two clear full-length tracks had been found on a muddy logging spur at the end of a trail through the salmonberries created by something very large. More tracks were found about a quarter mile along the spur. These had a nine-foot step and showed only the imprint of the forefoot. The tracks were exceptionally distinct in the mud. What appear to be pathologies of the head and base of the fifth metatarsal are evident along the outer edge of the cast of the complete foot. The cast of the half-track is truncated at a point that likewise approximates the inferred position of the midtarsal joint, the transverse line of flexion in the midfoot.

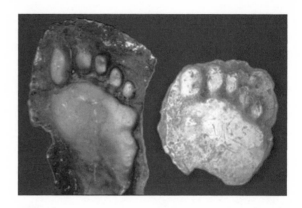

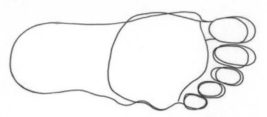

Casts made by Deputy Sheriff Dennis Heryford in Grays Harbor County, Washington. The cast on the right is another example of a half-track. Superimposition of the two outlines reveals the point of midfoot flexibility.

Even more examples have been identified. Perhaps the most intriguing case in point is present in the Patterson-Gimlin film. The film subject potentially affords a remarkable view of the plantar surface of the sasquatch foot, as well as an unobstructed perspective on several step cycles. In addition to a prominent elongated heel, a midtarsal flexibility is evident during the stance phase of gait, and considerable flexibility can be observed in the midfoot during the swing phase. A long line of deeply impressed footprints was left in the loamy sandbar at the film site. Patterson cast single examples of right and left feet. He selected the clearest, flattest, least distorted examples to be found. The casts are actually excellent representations of the anatomy of the static foot, but display little by way of dynamic features, nor does the single pair convey an obvious sense of animation that might be evident in a sequence of footprints. As a result, they made relatively little impact on the experts of the time, who regarded them as likely made by flat, carved fake feet.

Fortunately, Gimlin had covered a number of the footprints with slabs of bark pulled from the dead snags piled on the sandbar, shielding them from the rains that came during the following morning. Patterson and Gimlin didn't return to the site again, but two days later, on the following Monday, a Forest Service timber cruiser, Lyle Laverty, and his crew of seasonal employees visited the site. They had been camped near the junction of Notice Creek and Bluff Creek since early summer, preparing timber sales in the region. On the weekend following the filming, they were in

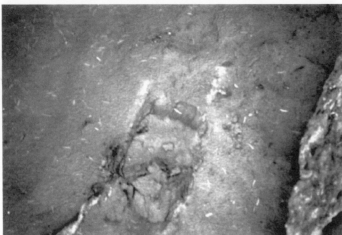

Photographs of footprints at the Patterson-Gimlin film site illustrating dynamic features. Note especially the pressure ridge particularly evident in the upper left footprint, subsequently among those cast by Bob Titmus. (Courtesy of Lyle Laverty)

Sequence of stills from the Patterson-Gimlin film exhibiting midtarsal flexion in the foot at push-off (Courtesy of Martin and Eric Dahinden)

town in Orleans and heard about the incident. On Monday they drove up to the site to have a look for themselves. They observed the footprints, and Laverty took some pictures of the most distinct prints. The dimpling effects of the rainfall can be seen in the photographs. Laverty followed the tracks along the sandbar for several hundred feet. Things seemed to be as Patterson and Gimlin have since described them. In contrast to Patterson's casts, Laverty's pictures depict examples of very dynamic footprints with features such as distinctive pressure ridges and discs in the midtarsal region. One pho-

tographed footprint in particular clearly exhibits a pronounced pressure ridge across the midfoot. These pictures proved to be extremely informative.

Nine days after the filming, Bob Titmus, the taxidermist mentioned earlier in connection with the Bluff Creek incidents in 1958, then in 1967 a Canadian resident, arrived at the site. He came prepared to cast the footprints, which were still clearly impressed in the sandbar, relatively protected by the bark, although the effects of the rain had left its mark before Gimlin was able to get them covered. In all, ten successive footprints were

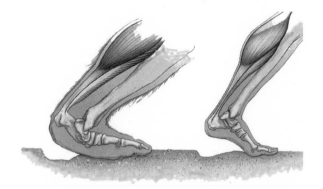

Contrast between the inferred midtarsal break in the sasquatch foot and the longitudinal arch of the human foot. Note the position of the resulting pressure ridge. Also notice the elongation of the sasquatch heel.

A human footprint in moist sand often leaves a pressure disc just behind the ball of the foot.

cast, providing a distinctly unique record of the potential variation present in a single sasquatch trackway. I have examined the original casts myself, and they are now accessioned at the Willow Creek–China Flat Museum collection. High quality silastic rubber molds of eight of the ten original casts, prepared by Grover Krantz, are now curated in the special collections of the Smithsonian. Animation of the foot is quite evident in this series of footprints. Three of the ten casts are of the very footprints photographed earlier by Laverty, providing exceptional corroborative documentation of these significant data.

What accounts for the distinctive half-tracks and pronounced pressure ridge that so distinguishes these tracks from human footprints? It is the flexible midfoot characteristic of the hominoid, or ape, foot. An ape's foot lacks a fixed longitudinal arch, and the joints at and below the ankle permit a greater range of motion than present in the human foot. The reconstructed sasquatch foot also lacks a fixed arch, as indicated by the clearest examples cast by Patterson himself, which are quite flat but certainly not featureless. In fact, the outline of the navicular can be readily identified on the inner side of the foot, just as it is evident in an ape footprint, or a very flat human footprint. In the typical human foot this bone is elevated in the apex of the arch, unless the foot is significantly pronated, or rolled inward. In contrast, the sasquatch foot retains the primitive apelike characteristic of a flat flexible midfoot. The range of motion permitted by the calcaneocuboid and talonavicular joints allows the heel to function in leverage and propulsion somewhat independently of the relatively prehensile function of the forefoot. In primates this coordinated flexion between the two seg-

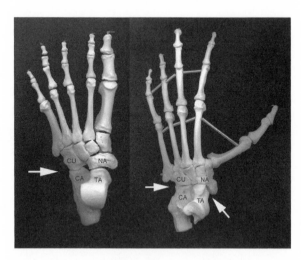

The left foot skeleton of a human foot (left) and a chimpanzee (right) indicating the position of the midtarsal or transversetarsal joints (arrows). Cu = cuboid, Ca = calcaneus, Na = navicular, Ta = talus.

ments of the foot is referred to as the "midtarsal break."

To appreciate the significance of this feature we must turn to the generalized ape foot. In the grasping foot of the ape, the great toe, or hallux, functions in opposition to the relatively long lateral digits in a pincer-like grip. This foot posture is especially evident when the ape is climbing on vertical or inclined tree trunks or branches—an ape specialization. The forefoot functions as a prehensile or grasping organ, maintaining a secure grip with its divergent big toe, while the hind foot provides leverage. The plantarflexors of the ankle elevate the heel as the power arm of a lever with its fulcrum at the midtarsal, or transverse tarsal joint.

The midtarsal break is also evident in the chimpanzee foot when it is walking on the ground. In studies of the pressures beneath the chimp foot, lifting the heel shifts the center of weight ahead of the midtarsal joint, especially beneath the cuboid bone. Generally the entire midfoot bears the weight rather

14-inch footprints in the foothills of the Blue Mountains of southeastern Washington. Note the pressure ridge. (Courtesy of Vance Orchard)

than concentrating it under a specialized ball of the foot. A similar pattern of weight distribution is implied by the dynamics of some examples of sasquatch footprints. The deepest part of the imprint lies just anterior to the midtarsal joint. If the ground is plastic enough, it may be pushed up behind the midfoot following the midtarsal break, creating a pressure ridge. I have identified repeated and geographically widespread instances of these pressure ridges in sasquatch footprints.

This midfoot flexibility also explains the relatively elongated heel attributed to the sasquatch foot. Critics of the Patterson-Gimlin film took exception to a protruding heel that suggested to them oversized fake feet worn by a "man in a fur suit." One skeptic went so far as to assert that the Achilles tendon appeared to attach far forward on the heel in an unnatural fashion, negating any mechanical advantage afforded by the elongated heel. These explanations are simply based on erroneous interpretations of the appearance of the subject or blatant ignorance of foot mechanics, and they can hardly

Two 14-inch footprints examined by the author in the foothills of the Blue Mountains of southeastern Washington in February 1996 displaying midtarsal pressure ridges

account for the dynamic actions and appearance of the feet evident in the film and the associated tracks.

Another interpretation based on sound principles of biomechanics would be that a means of increasing foot leverage needed by a bipedal animal with the large body weight of a sasquatch would be to increase the relative length of the heel. This trend is well documented in other apes of increasing body weight, from gibbons to gorillas. Indeed, if a series from chimp to lowland gorilla to the more terrestrial mountain gorilla, then on to sasquatch is considered, it can be seen that the sasquatch foot is a natural extension of the trends accompanying increased size and terrestriality, i.e., elongated and broadened heel, greater alignment of the great toe, and shortening of the remaining digits. Early hominids appear to have applied these principles prior to a stabilization and refinement of the longitudinal arch and associated adapatations to modern human walking and running. Once the transition to a fixed arch was made, humans modified the strategy significantly. Rather than lengthening the heel further, leverage was increased by shifting the pivot point, or fulcrum, from the midtarsus to the ball of the foot, and incorporating the midfoot segment into the power arm of the lever. Then the trend for heel elongation reversed itself in parallel with mechanical adaptations evident in other cursorial or running mammals.

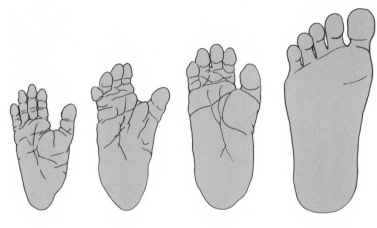

A series of feet—(left to right) chimpanzee, lowland gorilla, mountain gorilla, sasquatch—illustrating a trend for relatively shortened toes, aligned great toe, and elongated and broadened heel

The heel of modern humans became shorter to increase the speed ratio of the foot lever system, as endurance running became a primary human locomotor adaptation.

In the proposed sasquatch foot with an elongated heel, the gap between the Achilles tendon and the tibia, or shinbone, is wider. When the tendon is comparatively relaxed during the swing phase of gait, the heel appears to protrude in a nonhuman fashion. It was from a single poorly redrawn film frame depicting this situation that the misplaced attachment of the calcaneal tendon was inferred by the critic of the Patterson-Gimlin film subject. When the calf muscles are contracting and the tendon is taut during the stance phase, it obviously inserts on the proximal end of the calcaneus, and the heel no longer seems to protrude and the ankle appears unhumanly thick. In this regard the appearance of the Patterson-Gimlin film subject's heel is consistent not only with the dynamic signature of the accompanying footprints, but also with the bio-mechanical expectations for a foot supporting a giant bipedal ape, not to mention the observable action of appropriate muscles at key points in the step cycle.

An example of a sasquatch footprint exhibiting an apparent deformity may serve to highlight these distinctions of skeletal anatomy. These tracks were discovered near Bossburg, Washington, in 1969. The shenanigans that ensued from the incident have cast a pall of suspicion over the tracks. Kenneth Wylie, author and naturalist, alleged that it was confessed resident faker Ray Pickens who was responsible for the Bossburg tracks. Pickens appeared on television displaying his wares, an oversized pair of carved wooden fake feet attached to work boots. The devices were on par with the crude creations whittled by Rant Mullens. Furthermore, by his own public admission, Pickens's initial escapade with his concocted feet came long after the cripple foot tracks were originally discovered at Bossburg in December of 1969. He told reporters he first made his carved feet in February 1971.

In spite of the high jinks, the Bossburg tracks remain a remarkable set of footprints deserving of consideration on their own merits. Two interested scientists, John Napier and Grover Krantz (not to mention a number of orthopedists and podiatrists that I have conferred with) took the footprints quite seriously. The malformed right foot caused Napier to reflect, "It is very difficult to conceive of a hoaxer so subtle, so knowledgable—and so sick—who would deliberately fake a footprint of this na-ture. I suppose it is possible, but it is so unlikely

Still from a television news report in which Ray Pickens displayed a false foot he used to hoax footprints

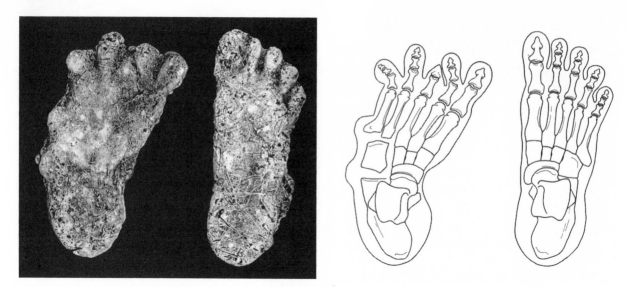

Paired casts of the "crippled" sasquatch at Bossburg, Washington. The author has inferred the skeletal proportions (differing somewhat from the reconstruction of Krantz).

that I am prepared to discount it." Napier examined only a very limited sample of footprint casts and photos, including the Bossburg casts. On this basis he inferred two distinct types of footprints, which he referred to as the "hourglass" and the "human" types. The hourglass type was distinguished by the waisted appearance of the shank of the foot. They also displayed toes that were more equal in size, separated from the sole by a ridge of soil. Napier made a further point about an observation that the heel impresses more deeply on the outer edge of the heel, the gait is pigeon-toed and the oblique alignment of the toes indicates a push off from the outer side of the foot. Although impressed by both types, he found it implausible that there were two varieties of Bigfoot and eventually cast doubt upon, then rejected the hourglass type. It appears that the principal example of the human type was the footprints from Bossburg. Since this assessment has been cited in support of the contention that the Bossburg tracks are merely enlarged facsimilies of a malformed human foot copied from an orthopedics textbook, it is worthwhile to examine Napier's criteria more closely.

First of all, Napier placed a great deal of significance on the relative size of the great toe. He provided a table of values for the big toe width index [(big toe width × 100)/combined width of all toes] for a sample that comprised a mere seven sasquatch casts and six humans (all measurements derived from photographs of footprints). The Bossburg big toe index was reported as 35.7, compared to average values of 23.3 (range of 18.2–28.7) for the five remaining hourglass types, and 32.8 (range of 31.1–36.5) for the human type. In an attempt to replicate Napier's measurements for the Bossburg footprint casts I arrived at an index value of 25, well within the hourglass type's range and outside that for human footprints. The confusion may have ensued from the degree of splaying of the toes, or from the ambiguity in the boundary between the first and second toe in the particular Bossburg cast Napier was examining. If one allows for the appropriate width of the second toe, assuming it is minimally as wide as the third toe, then the big toe is smaller than inferred by Napier. This is borne out by reference to the malformed foot. The lateral toes are misshapen and apparently enlarged, but the great toe seems unaffected. It seems unlikely that it would become smaller in width, while the lateral toes were hypertrophied. In fact its width is a good match for the inferred width of the big toe on the opposite foot. Hence the principal criterion for the classification of the Bossburg tracts as *human type* cannot be supported.

The remaining distinctions have more to do with the nature of the substrate in which the footprint was left, than with the feet themselves. In a thin veneer of dust on a hardpacked logging road, a gap is more evident between the toe pads and the sole than when the same foot leaves a deeper imprint in soft soil. The same gap is suggested by a wet human footprint on an unyielding cement surface. This is exaggerated by the relatively longer toes apparent in the sasquatch footprints. When they are curled in flexion the ridge is more pronounced. The toes of the Bossburg tracks are flexed to a considerably lesser degree than the examples cited as hourglass types. They are also aligned obliquely, similar to the hourglass type, and indeed the shank of the foot is waisted, flaring to a broad heel (which Napier inaccurately rendered in outline in his figure). There is no contradiction between the sasquatch footprints, no distinction into two types. The Bossburg tracks do not conform to a human pattern, but are as distinct from human footprints as any of Napier's hourglass types.

Krantz examined the "cripple foot" tracks in the ground firsthand and they embodied for him the first convincing evidence for the reality of sasquatch. The deformity exhibits an anatomy consistent with the general condition referred to as *pes cavus,* specifically a type of skewfoot. Its unilateral manifestation is suggestive that it resulted from a spinal cord lesion, rather than a congenital deformity. Regardless of the epidemiology, from the pathology Krantz inferred the position of the joints disposed on the outside edge of the foot. He considered the protrusions presumably caused by hypertrophied connective tissue to be associated with the deranged joints proximal and distal to the cuboid bone. These prominent bunionettes on the outside margin of the foot, resulting from the inward twisting of the forefoot, are positioned more distally than would be the case in a human foot suffering from a similar condition, but are in accord with the proportions of an elongated heel segment and the position of the midtarsal joint as inferred from the pressure ridges and half-tracks already examined. The resulting skeletal proportions and joint positions are remarkably consistent with the model of sasquatch foot morphology suggested here based on independent examples of footprints.

Skeptics have argued that Krantz's reconstruction does not hold up because the ankle joint, the tibiotalar joint, does not contact the ground and therefore cannot be inferred from a footprint. They argue further that no skeletal feature can be inferred from the impression of soft tissue and that the protrusions on the lateral side of the cripple foot could be interpreted in any number of ways. Such comments expose a lack of understanding of the correlated nature of foot anatomy and no appreciation for the features discernable in footprints. Krantz's placement of the tibiotalar joint was based not only on the cripple foot casts, but derives from his observations of the Patterson-Gimlin film subject as well. The film subject displayed ankles that were quite thick from front to back. The placement of the tibia, or shinbone, is just beneath the skin at the front border of the leg. Therefore reasonable placement of its long axis is evident, even allowing for body hair. Furthermore, the relationship of the talus to the anterior calcaneus is quite consistent, excepting highly specialized leapers such as tarsiers. The notable variation comes in the relative length of the calcaneal tuberosity, i.e., the proximal length of the heel. This elongated heel, to which is attached the Achilles tendon, is what lends the ankle its apparent thickness in the film subject. The fact that the relative length of the calcaneus, attested to by its appearance in the film subject, corresponds so closely with the relative placement of the proximal bunionette in the cripple foot, seems too fine a point to be sheer coincidence, especially when it correlates with the multiple independent examples of midfoot flexibility, midfoot pressure ridges, and half-tracks.

The allegation that no skeletal feature can be inferred from the impression of soft tissue can be readily refuted. I sampled a set of weight-bearing footprints from a large and diverse class of university students (a sample of more than fifty students). Their soles were dusted with fingerprint powder, after which they stepped onto a strip of lifting tape, which was then applied to card stock. In the majority of these, the prominence of the tuberosity of the fifth metatarsal was evident as a convexity on the lateral edge of the foot. Proximal to it lies the cuboid, from which the position of the calcaneocuboid joint can be approximated. In those fewer individuals with flat pronated feet, the tuberosity of the navicular was evident as a convexity on the medial side of the foot. The talonavicular joint is just proximal to it. With this information it is a simple matter to approximate the position of the transverse tarsal joint from the footprint. The positions of the heads of the first and fifth metatarsals are indicated by the widest breadth across the forefoot immediately proximal to the toes. Flexion creases on the toes correspond in typical fashion with the interphalangeal joints. The development of the apical pads of the digits can also imply aspects of toe proportion. In reality, a great deal about the foot skeleton *can* be deduced from the features of a footprint.

Deformities and misalignments of the toes also permit inferences to be drawn about the positions of joints and thereby the relative toe lengths. Together with other examples of pronounced toe mobility as evidenced in the variability of toe position from footprint to footprint, they indicate that the sasquatch toes are relatively longer and more mobile than human toes, even those of individuals with feet not usually confined in

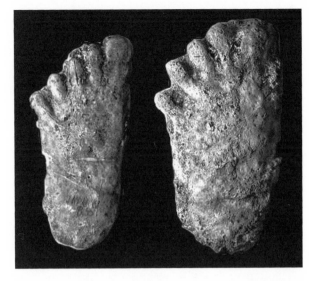

Two casts from southeastern Washington exhibiting obviously long toes. On the left, a cast made by Wes Sumerlin near Indian Springs in 1988; on the right, a cast made by Paul Freeman south of the Mill Creek Watershed in 1987. The casts represent the same individual.

shoes. In some instances the footprint indicates long toes that are sharply curled in a flexed posture, leaving an undisturbed ridge of soil behind the toe pads. Gripping in this fashion the toes are also adducted, or pressed in against each other, side-to-side. This packed position may contribute to the squarish appearance of the toes. With only the toe tips in contact with the ground they look much like "peas in a pod," as noted

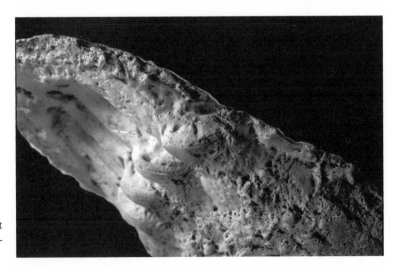

Profile of the fifth toe of the half-track cast by the author outside of Walla Walla, Washington, in 1996

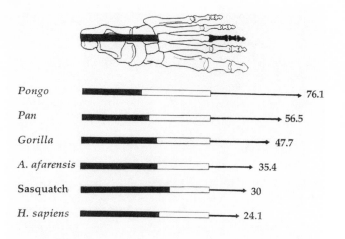

Pongo ————————————————→ 76.1

Pan ———————————→ 56.5

Gorilla ——————————→ 47.7

A. afarensis ———————→ 35.4

Sasquatch ——————→ 30

H. sapiens —————→ 24.1

Comparison of the relative mean toe length (third digit) in a sample of hominoids. The combined length of the tarsus and the third metatarsal is standardized to a single unit, and the values to the right express the length of the toe as a percentage of the unit length. Note that the sasquatch are approximately intermediate to those of modern humans and an early bipedal hominid (modified from Tim White and Gen Suwa).

by Napier. In other instances the toes are fully extended and give a truer impression of their absolute lengths and proportions. In this position, they are sometimes abducted, or splayed apart from each other, occasionally with mud extruded through the gaps between the toes. Examining one's own fingers can provide a dramatic illustration of these contrasting situations. When extended, the fingers are rather disparate in length and apparent size, but curl them toward the palm and suddenly they appear nearly equal in length with rather uniform tips lined up like peas in a pod.

Among the casts that I made in 1996 in southeastern Washington was one in which the toes had deeply impressed in the soft mud and the first and fifth digits splayed against the sidewall of the track. The result was a side profile of these flexed marginal toes, the first instance of such that I am aware of. The least toe on these nearly 14-inch tracks is over 3 inches long. In other words, the toes of this sasquatch foot would conceivably have as much prehensile, or grasping ability as a human hand (excluding the opposition of the thumb). Given the greater muscle mass of the lower leg of the sasquatch, the grip strength of the toes would be many times that of a man's fingers. Expressed as a percentage of the combined lengths of the hindfoot and midfoot, the sasquatch toes are intermediate in relative length between those of a human and an early bipedal hominid, *Australopithecus afarensis*. Allowing for the relative elongation of the hind foot segment of the much larger sasquatch, the toe percentage is actually much closer to the early hominid value.

The issue of toe length is related to another occasional feature that has drawn comment, and was noted by Napier as a feature of his "hourglass" type. This is the apparent "split ball" that was particularly evident in a couple of examples of footprints in northern California and was grossly exaggerated in the fake feet carved by Ray Wallace. Rather than being a hallmark of the sasquatch foot, as mistakenly asserted by some, only a limited number of individuals exhibited this feature to a pronounced degree—the 15-inch and 13-inch tracks found on the Blue Creek Mountain Road in 1967, and to a much lesser degree the original 17-inch individual. Dr. Grover Krantz had concluded that assuming the sasquatch midfoot skeleton was comparable to a human's, the tremendous weight of the sasquatch would dictate toes and metatarsals relatively shorter than a human's. In a flat foot, both the proximal as well as the distal ends of the hallucial metatarsal would be pressed to the ground, creating a second ball around the proximal joint of that bone, hence a split or "double ball." However, examination of numerous casts produced only rare additional examples of the double ball feature, but presented many examples of long splayed toes. The resolution to this apparent contradiction came one evening while I was indulging my wife with a foot massage as we watched a late movie together. My wife has very flexible toes, which she can actually curl beneath her foot, resting upon the back of the toe "knuckles" when she is sitting, a trait she has passed along to most of our boys. She also has a relatively thick sole pad, which extends nearly to the first interphalangeal joint, giving her toes a quite stubby appearance when viewed from below, i.e., from the perspective of a footprint. As I massaged her foot, I rubbed my palm over the tops of her toes, pressing them into flexion. As I did, a distinct crease appeared across the ball of her

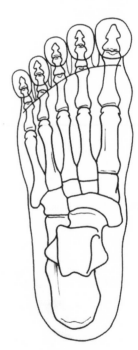

foot that extended nearly its full width. This flexion crease corresponded to the position of the metatarsalphalangeal joint around which the ball of the foot is situated. The same relationship can be more handily observed by flexing one's fingers as when waving "bye-bye." The action emphasizes a crease running across the palm that corresponds to the position of the proximal knuckle joints. The same action applied to my foot produced no obvious crease, as my sole pad is much thinner and does not extend as far beyond the bases of my toes as my wife's, nor do my toes exhibit a comparable range of flexion.

In fact, all humans are born with a flexion crease across the ball of their foot, as can be readily confirmed by referring to the inked footprint on one's birth certificate. As development of the foot continues and the ball of the foot fills out and forms more fully, the crease is generally lost. It may persist to varying degrees in persons with hypermobile flat feet, however. The presence of a split ball to varying degrees in selected sasquatch tracks indicates a thick

Even the tracks from the Patterson-Gimlin film site, which are often described as showing stubby toes, actually indicate relatively long toes, obscured by an extension of the sole pad as indicated in this inferred reconstruction.

extensive sole pad in those particular individuals, which bunches up when the long toes are sharply flexed. In reconstructing the skeletal proportions of the footprint from the Patterson-Gimlin film site, I placed the metatarsalphalangeal joints as indicated by the faint flexion crease. This produced toes of a length generally consistent with other sasquatch tracks in which the toes only appeared to be much longer. Recent enhancements of the Patterson-Gimlin film, using digital color-channel separation technologies, have provided never-before-seen clarity of detail of the film subject's anatomy. I was especially intrigued by a stabilized sequence of frames that provided a view of the dorsiflexion of the toes near the end of the swing phase of a step. The toes briefly angle upward just before the foot makes contact with the ground. They are noticably long and exhibit an excellent proportional match to the reconstructed lengths inferred from the plantar flexion crease. I have likewise redrawn the skeletal proportions of the Bossburg cripple foot to reflect the longer toe proportions, which produces better agreement with the soft tissue anatomy evident in those footprint casts. Details of the longer toes were confirmed, corroborating my revised reconstruction, when I was afforded the opportunity to examine and photograph the *original* casts, courtesy of Chris Murphy. Therefore, the "split ball" feature appears to have been a rather restricted regional idiosyncrasy that is not characteristic of sasquatch tracks generally, yet is a consistent expression of the expected variation within the hominid-based model of the sasquatch foot.

The overall dynamic signature of the sasquatch footprints concurs with numerous eyewitness accounts noting the smoothness of gait exhibited by the sasquatch. For example, one witness observed, "It seemed to glide or float as it moved." Absent is the head-bobbing or vertical oscillation characteristic of the stiff-legged human gait. The compliant gait on flexed knees and hips not only reduces the peak impact forces under the feet, but also avoids the focused concentration of pressure under the heel and ball of the foot. Recall that these pressures would be disproportionate in a sasquatch-sized biped. Instead, the broad foot is placed flat on the ground, distributing weight more evenly across a large surface area and throughout a thick sole pad. The compliant gait also increases the period of double support, when weight is borne over both feet, simultaneously reducing the duration that an individual foot must bear the brunt of total body weight.

By contrast, human walking is characterized by an extended stiff-legged striding gait with distinct heel-strike and toe-off phases. Upward bending of the toes, placing elevated bending stresses on them, marks the end of the stance phase. Double support is minimized in order to maximize step length, resulting in the push-off coming more from the toes rather than from the forefoot. These bending stresses exerted on the toes are reduced by selection for relatively shorter straighter toes that participate in traction and propulsion at the sacrifice of prehensile capacities. Efficiency and economy of muscle action during endurance walking and running are maximized by reduced mobility in the ankle and hind-foot joints, limiting movements toward simple flexion and extension, a fixed longitudinal arch, which provides a stable platform of support and increased leverage, and elongation of the Achilles tendon for speed of flexion and elastic storage of kinetic energy.

These distinctive qualities of modern human walking may have been relatively recent evolutionary innovations. After the initial transition to habitual bipedalism, perhaps as early as 7 million years ago, the hominoid grasp-climb adaptation of the foot was compromised by moderate shortening of the lateral toes and reduction in the range of abduction, or divergence, of the medial toe, the hallux. The first direct evidence of early hominid bipedalism came with the discovery of fossilized footprints in petrified volcanic ash beds 3.7 million years ago at the site of Laetoli, in East Africa.

The Laetoli footprints exhibit these modifications to the prehensile portion of the foot to an intermediate degree. The big toe is considerably aligned with the remaining toes, but all are comparatively long.

There has been continuing debate over the extent to which the Laetoli hominids display modern human foot morphology. Some have argued that the footprints imply a foot essentially modern in all aspects, while others have pointed out features that indicate retention of more apelike characteristics of the foot. I first drew attention to what might be a midtarsal pressure ridge evident in a number of the fossil footprints, which suggested the retention of midfoot flexibility. This is remarkably reminiscent of the comparable feature in the sasquatch tracks. In a depiction of a reconstruction of the *A. afarensis* foot skeleton superimposed upon a Laetoli footprint, this feature can be seen to lie immediately proximal to the position of the reconstructed midtarsal joint. Some have suggested this feature is possibly the result of termite burrowing or an excavation artifact; however, the repeated and consistent position of the feature, combined with other

Cast of a 3.7–million-year-old Laetoli footprint of an early bipedal hominid, exhibiting a pressure ridge (arrow), a product of midfoot flexibility

Fossilized footprint of an 400,000-year-old hominid found in the south of France at the site of Terra Amata (Courtesy of Charles DeLumley)

indications within the footprints of a transverse axis of foot flexion, indicate its interpretation as a pressure ridge is justified.

An exceptional example of a midtarsal pressure release is evident in a particularly well-preserved Laetoli footprint (G1-26), indicated by plastic flow of the wet ash pushed up behind the midtarsus. Careful examination of this feature excludes the possibility of it resulting from exfoliation of successive layers of ash or other excavation artifact. The leading edge of the flow is continuously rounded and unbroken, and has the same appearance as the ash extrusion between the first and second toes of another very clear print (G1-36).

Recent analyses of fossil hominid foot skeletons also indicate midfoot flexibility in australopithecine, and perhaps early *Homo* feet. At some point thereafter, in the evolution of modern human foot morphology, changes occurred to stabilize the foot platform, increase mechanical advantage of ankle flexors, and improve efficiency and economy in long distance, endurance walking and running. Determining the timing and pattern of the evolution of these characteristics has remained a challenge due to the paucity of fossilized footprints or foot skeletons from the period spanning 2.0–0.5 mya. Two potentially critical specimens have remained largely unpublished. The first of these is the footprint at the Terra Amata site, in southern France. The site dates to approximately 400,000 years ago. The single published photo of this footprint suggests a midtarsal pressure ridge and a lack of a well-developed longitudinal arch, as well as the lack of a differentiated ball. This seems to agree with the observation that a single metatarsal of the big toe attributed to *Homo erectus* is remarkable in its lack of robustness. The gracile proportions of the bone combined with its distal joint morphology suggest it was part of a foot that lacked a well-developed ball. The footprint also appears relatively long for its breadth. This and relative position of the apparent pressure ridge suggests a degree of heel elongation. This would increase the leverage of the ankle flexors in response to dramatically increased body mass in *Homo erectus* over previous hominids. Although not as significant as the proposed gigantism in sasquatch, the parallel in biomechanical response is very suggestive.

The second specimen of interest is the nearly complete foot skeleton of the Jinniushan hominid from a site in China dating to just less than 200,000 years ago. A photograph of the skeleton has appeared in a popular Chinese magazine, but detailed descriptions or analyses of the foot skeleton have yet to be published. Features of the foot skeleton visible in the magazine photo suggest that stabilization of the transverse tarsal joint had occurred by that time. Specifically, the shapes of the joints of the midtarsus reveal that they could lock in a stable position to support a longitudinal arch. Also the bones of the big toe are large and thick, indicating the importance of toeing-off in this species. Interestingly, it appears that the heel remains relatively elongated.

Therefore, based on this admittedly limited assessment, it emerges that the transition to the modern foot form, characterized foremost by the longitudinal arch and well-developed ball and big toe, may have occurred only as recently as less than 200,000 years ago. In that case, it should now be recognized that the majority of the history of hominid bipedalism reveals a sustained and apparently successful strategy of locomoting bipedally on flat flexible feet. Modern human foot morphology was perhaps a surprisingly recent evolutionary innovation. The stabilization of the midfoot and associated modifications, especially to the distal foot, i.e., shortening of the toes, enlargement of the big toe, development of the ball, as well as shortening of the heel, were relatively recent innovations that marked a shift in hominid locomotor adaptation to a lighter skeleton combined with endurance walking and running.

The sasquatch appear to have adapted to bipedal locomotion by employing a compliant gait on flat flexible feet. A significant degree of prehension has been preserved in the toes by retaining the uncoupling of the propulsive function of the hind foot from the forefoot by means of the midtarsal break. The toes are spared the peak bending forces experienced by human toes at toe-off by means of the compliant gait with its extended period of double support (the interval of a gait during which both limbs are in contact with ground and support body weight) and push-off from the front half of the foot. All in all, this would be an efficient strategy for a giant terrestrial bipedal ape. In fact, it is an extension, most likely in parallel, to the very strategy that marked the mode of bipedalism used by early hominids. The sasquatch foot and footprints exhibit intriguing parallels to the morphology of the hominid foot, maintaining aspects of the longstanding bipedal hominid strategy, while modifying it, e.g., increased breadth and further lengthening of heel, to accommodate its gigantic body proportions.

The inferred architecture of the sasquatch foot is not only well documented, but seems well suited to the physical aspects of the terrain of its purported range. The retention of somewhat prehensile toes, combined with increased leverage of the heel, give it an advantage negotiating the steep and uneven mountainous forest landscapes of North America. The locomotor adaptation of an organism is a major element in defining its niche. The conformity of the inferred sasquatch locomotion to an overall hominoid/early hominid framework, and the anatomical distinctions correlated to its environmental specializations are plausible and compelling arguments for a real animal. Could such coherence of form and function emerge from an unrelated series of spurious counterfeits?

LINE UPON LINE: DERMATOGLYPHICS

A distinctive characteristic defining the order primates is the presence of hairless friction skin on the palms of the hands and soles of the feet. Friction skin is an adaptation for surer grip in an arboreal setting, among the narrow and unsteady branches of the forest canopy. The papillary ridges in the dermis of the palms and soles produce lines of ridges separated by furrows in the overlying epidermis, which are called dermatoglyphics. The configuration of the ridges displays distinctive characteristics, such as ending ridges, short ridges, and bifurcations. The particular patterns of ridge flow combined with a unique set of distinguishing characteristics give each individual a set of distinctive fingerprints that can be used for purposes of identification. The precise interactions of genetics and environment that combine to establish these patterns during development is not understood. What is sometimes not realized is that the ridged skin not only extends across the palms of the hands, but also covers the undersurface of the toes and soles of the feet. The core patterns, arches, loops, and whorls that are characteristic of fingerprints are situated below the tips of curled toes and usually go unregistered in a footprint.

Along the summit of the skin ridges, at variable intervals, are small depressions, pits that mark the openings of sweat glands. These eccrine sweat

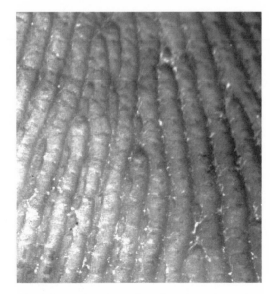

Close-up of the skin ridges found on the palms and soles of primates. Note the pores of sweat glands scattered along the crest of the ridges.

glands produce a watery sweat that moistens the soles and palms, thus increasing the friction coefficient and improving grip. This principle is seen in action every time a lumberjack spits into his hands before taking up the axe.

Only under appropriate soil conditions, that is, when the soil particles are fine enough, can ridge detail be transferred from the sole of the foot to the surface of a footprint. In some places in western North America, the soil contains large amounts of loess, a type of calcareous silt. Loess is a windblown dust derived from rocks pulverized by the grinding actions of advancing glaciers during the Ice Age. It has the consistency of talcum powder. There are particularly significant deposits of loess in eastern Washington and southern Idaho. Another source of fine soil elements are the volcanoes of the Cascade mountain range. Deposits of fine volcanic ash, or tephra, have been recorded throughout the West and in places are exposed in road cuts or deposited as silt along streambeds. On a smaller scale, unimproved logging or Forest Service roads can become dust beds in dry weather as the traffic pulverizes the dirt into extremely fine dust. The powdery dust picks up fine details of the skin, including ridge detail.

Ridge detail in a footprint is a rather transient feature and may persist for only a matter of hours, depending on the soil and weather conditions. Gravity, wind, drying, or moisture relentlessly acts upon the details of footprints to level them out. If the soil is moist and the weather wet, this is accomplished very rapidly. I conducted a simple experiment to demonstrate this to myself. The soil in the hills surrounding Pocatello, Idaho, contains a large amount of loess. On a wet overcast day I went to a clear spot in the garden and with bare feet stepped off a series of clear footprints, each displaying extensive dermatoglyphics. These were most visible under the toe tips and beneath the forefoot. I immediately made a cast of the first footprint. The cast retained the majority of the transferred friction skin detail. Four hours later I examined and cast the next footprint in the row. Less than half of the ridge detail remained and transferred to the cast. Another four hours later, traces of detail could be seen only at the edges of the third footprint and about the toe stems. Interestingly, these are the same areas of a footprint in which ridge detail is most likely to be present when discernable on a sasquatch footprint cast. If the soil is dry and dusty, passing vehicular traffic eventually inundates the footprints with clouds of dust that quickly obliterate the unconsolidated fine detail. The transient nature of the dermal ridges and their differential persistence in certain areas of the foot compared to others sometimes goes unappreciated. Instead the patterns of distribution are attributed exclusively to wear and abbration of the plantar skin, leading to some unsupported criticisms of the ridge details.

The skin of the sole of the foot, the plantar skin, gives rise to a keratonized layer that is substantially thicker than anywhere else on the body. This layer becomes even thicker when the feet go a lifetime without shoes and may increase by as much as 1 cm in thickness. Such thickening, combined with the abrasion that naturally accompanies unshod walking and running, may wear down the ridge detail to some degree, especially on areas of concentrated pressure beneath the foot. Therefore the presence of evident ridge detail on a sasquatch foot may also vary with the age and health of the individual and may also vary seasonally with its ranging behavior.

Given these individual variations, as well as the specific prerequisite soil conditions and the transient nature of the ridge detail even under ideal conditions, it is not surprising that dermatoglyphics are rarely identified in sasquatch tracks. But if dermatoglyphics are present at the time a cast is made they can indeed transfer to the surface of the cast, if the casting medium is prepared and poured properly. The existing examples represent an extremely small sample.

Close-up of one of the casts from southeastern Washington, on which Dr. Grover Krantz recognized skin ridge detail (Courtesy of Grover Krantz)

Dr. Grover Krantz was the first to draw widespread attention to the presence of ridge detail in several sasquatch footprint casts from the Blue Mountains of southeastern Washington in 1982. In some instances the preserved resolution of detail was such that individual sweat pores were apparent and could be distinguished from artifacts caused by trapped air bubbles in the plaster. Krantz's finds received a cool reception from the anthropological community. Few academics would take a serious look at the dermatoglyphics given their declared source. Some were quick to point out ways in which a rubber mold of a human or ape foot could readily be enlarged in solvents producing a "big foot" with enlarged ridge detail. Of course this proposition ignores the distinctive anatomical details and proportions of the footprints that are absent in either a human foot or ape foot. It is merely a simplistic rationalization for a complex occurence.

The interpretation of sweat pores was drawn into question and dismissed as trapped air bubbles in the plaster, even though the bubble artifacts with their sharp edges could be readily distinguished from the rounded pits resembling the appearance and distribution of sweat pores. John Berry, Fellow of the Fingerprint Society (FFS), and Stephen Haylock, FFS, published a commentary in *Fingerprint Whorld* on the casts from southeastern Washington. They confirmed the presence of sweat pores and confirmed that these could readily be distinguished from trapped air bubbles in the casting medium.

Dr. Krantz summarized the reactions to the dermatoglyphics as follows: "The reactions of two major groups are significantly different. In the fields of physical anthropology and zoology, most of the responses have ranged from mild interest to flat denial that they could be real. Among fingerprint experts and pathologists, the responses have ranged from mild interest to enthusiastic acceptance."

For example, Edward Palma, fingerprint examiner for the Laramie County Sheriff's Department, Cheyenne, Wyoming, concluded after examining latex lifts and the original plaster casts that they represent footprints of a living higher primate of an unknown species. He discounted the suggestion that they could be enlarged human feet because of the disproportionate breadth of the foot compared to a human, which was nonetheless supported by the ridge pattern. Palma was able to trace the ridge pattern over the entire breadth of the forefoot, finding appropriate landmarks in their respective positions with intervening ridges flowing in proper directions. He could not see how the patterns could have been patched together from smaller parts that were copied from skin of a known primate. Beyond the overall patterns, the detailed structure of the ridges conformed to real friction skin. The evident sweat pores were aligned and spaced as expected and distinct from the occasional bubble in the plaster. Finally he confided, "I began this investigation with the goal of showing how these prints were, or might have been, faked. All evidence now tells me that any faking would be impossible."

Benny Kling, Law Enforcement Academy Instructor, Douglas, Wyoming, likewise examined inked latex lifts and the original casts and drew the same conclusion that the casts exhibit characteristics of real friction skin from a higher primate. He added a note on the apparent mirrored symmetry of the right and left feet. He pointed out breakdown and smoothing of the ridges under weight-bearing areas, as might be expected of a living foot, are near mirror images; some dysplasia is indicated in the areas where it could be expected, smoothing by wear shows on the weight-bearing areas. In summary, he said, "This kind of print could not have been made by a human foot, or that of any known animal. It could not have been manufactured by a hoaxer; the design is too dermatoglyphically correct, and the engraving job would be beyond the capabilities of the best forger."

Track casts in northern Idaho, with a close-up of the resin cast, displaying skin ridge detail (Courtesy of Jim McLeod)

Although the dermatoglyphics in the casts from Walla Walla drew the most attention, they were not the only, or even the first, casts to bear such indications of skin ridge details. I subsequently identified a number of additional casts of sasquatch footprints that also appeared to display patches of ridge detail. An earlier example comes from Blanchard, Idaho, in 1977, and had been investigated by Dr. James McLeod, of North Idaho College, and student John Witherow. A couple witnessed a large hairy figure making its way up the mountainside and across a dirt road on Mt. Spokane. Upon investigation, reserve deputy and experienced tracker Wayne Rasmussen discovered and cast two clear 17.5-inch footprints where the sasquatch had stepped in the moist silty clay of a dissipated puddle. He subsequently tracked the animal for a considerable distance. The tracks had "lines" in them, which resembled dermatoglyphics. Rasmussen made two casts, one of plaster and one of acrylic resin, preserving the dermal ridge detail.

I also had the opportunity to carefully examine the original cast material that taxidermist Bob Titmus made on two occasions in 1963 near Hyampom, California. These were made once in fine ash from a slash burn and later in wet soil and mud. One of the casts preserved traces of coarse skin ridge detail along the inside margin of the footprint where the sole pad had expanded under weight and undercut the edge of the footprint. These tracks are also of interest because they indicate the position of the joint at the base of the big toe, the hallucial metatarsophalangeal joint. One cast preserves a pressure ridge just behind the point of joint flexion. A second cast indicates that the foot stepped on a rock and the joint flexed to accommodate it. The lateral toes, especially the third and the fourth, curled sharply over the stone and impressed deeply in front of it. Both of these features correspond to the location of a subtle flexion crease. The large footprints bear a striking resemblance, based on shape and

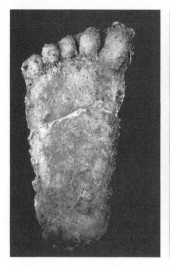

Pair of casts from Hyampom, California, 1963, from Bob Titmus's collection. On the right is close-up detail on the medial side of the cast where the sole pad has expanded and undercut the edge of the print. Indications of ridge detail are present.

proportions, to the large track originally cast at Bluff Creek in 1958 and are very likely to have been left by the same individual.

I made mention of some of these additional examples of skin ridge detail in a television interview that was watched by Officer Jimmy Chilcutt, a crime scene investigator and latent fingerprint examiner with the Conroe, Texas, Police Department. Officer Chilcutt also has extensive experience with nonhuman primate dermatoglyphics, an exceptional combination of expertise. He has printed hundreds of primates at zoos and research centers across the country, including the Yerkes Primate Center.

Chilcutt was intrigued by the possibility of dermatoglyphics on alleged sasquatch footprint casts and immediately contacted me and arranged to visit my lab and examine my collection of casts. I introduced him to the casts and then left him to examine them alone without any input from me. After several hours of surveying the material, Chilcutt's attention was focused on a few particular casts. These definitely exhibited dermatoglyphic features, but of a texture (ridge spacing and width) and flow pattern that were unlike what he was familiar with after many years of examining human and nonhuman primate finger

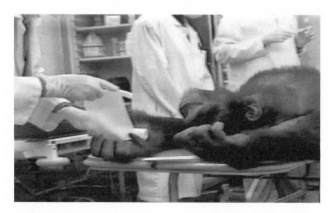

Officer Jimmy Chilcutt lifting a dusted palm print off an anesthetized chimp (Courtesy of Jimmy Chilcutt)

and palm prints. The ridges were on average twice as wide as typical human ridges, and where the human sole generally has ridges that run transversely across the width of the foot, ending perpendicular to the edge of the foot, the ridges on the margins of the sasquatch casts tended to lay parallel to the edge of the foot and generally run more-or-less lengthwise along the axis of the foot.

Palm print of a gorilla (Courtesy of Jimmy Chilcutt)

What most impressed Officer Chilcutt were multiple examples of healed scars that appeared on a particular pair of casts from the Blue Mountains in southeastern Washington, where the soil has a high content of loess. Dr. Krantz had previously referred to these casts as "Wrinkle Foot" due to the extensive indications of coarse dermatoglyphics. The deep, clear footprints were found in wet mud and preserve much detail of the skin surface. Chilcutt reasoned, "If this animal is walking through the wilderness, he's bound to come across rock and rough terrain that will cut the bottom of his foot. As the wound heals, the ridges curl inward toward the scar."

One cast in my assemblage came from a road on Onion Mountain near the Blue Creek Mountain Road, and was made in 1967, just a short time before the Patterson-Gimlin film was taken. It was the 13-inch track that exhibited coarse ridges about the margins of the pour. Krantz had assumed this was simply a pouring artifact, as had I. Chilcutt's attention had also been drawn to this specimen, because the coarse texture of the ridges was comparable to those on Wrinkle Foot from Walla Walla, Washington, and the ridges exhibited characteristics that suggested they were dermatoglyphics rather than some spontaneous artifact. The conditions where the tracks were laid down had been conducive to transfer of ridge detail. The logging road was covered with a layer of fine dust that had been dampened by a light rain. Green noted the extreme clarity of the imprints at the time of their discovery, "It even appeared to show the texture of the skin on the bottom of the foot, grooved in tiny lines running the length of the print." But every time a construction

The author and Jimmy Chilcutt examining ridge detail on casts of sasquatch footprints (Courtesy of Helen Hardy)

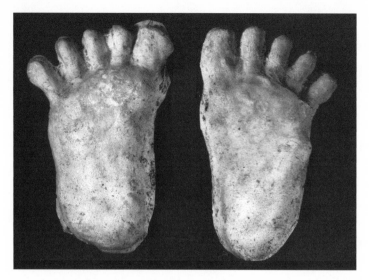

Paired casts of 13-inch footprints discovered by Paul Freeman in southeastern Washington, near Table Springs, along the Walla Walla River

vehicle would rumble past, a heavy cloud of dust was stirred up, only to settle across the tracks, quickly obscuring details.

The question arose whether these features might be an artifact of creating the cast in the fine dust. It was proposed that without consolidation of the delicate features through application of a fixative, the ridge detail was obliterated at the center of the pour under the weight of the plaster. Only as the plaster spread out and thinned at the margins were the ridges in the unconsolidated dust preserved. This phenomenon was repeated experimentally in the lab with human footprints left in fine loess and cast. Similar patterns of ridge preservation were replicated. Under the extremely dry conditions of the fine loess, which tended to wick the water away from plaster quite rapidly, pouring artifacts were occassionally produced that superficially resembled thick parallel ridges. These artifacts did not consistently exhibit uniform width or other fine details characteristic of dermatoglyphics; nor could these conditions account for ridge detail in casts poured in wet silt or mud.

Taped lift of the skin ridge detail of the cast from Table Springs

I later identified an additional 13-inch cast in Krantz's collection from the same site. Upon examining it, Chilcutt confirmed that it likewise displayed similar coarse ridge detail, although fainter, probably due to inundation by settling dust prior to casting. This observation affirmed his conclusion that these represented natural dermatoglyphics rather than pouring artifacts, or else one might expect the clarity of the ridge detail to be comparble in so far as the pouring technique was similar. Questions still remain concerning the possible occurrence of pouring artifact under hot, exceptionally dry conditions and further experimentation is needed. This challenge has been taken on by an amatuer investigator, Matt Crowley, whose preliminary results raise questions specifically about the interpretation of the Onion Mountain cast features as dermatoglyphics.

Chilcutt spent another couple of days scrutinizing the casts and comparing them with representative ape and human inked prints that he had brought along from his collection. During a break for lunch, Chilcutt broke a reflective moment of silence by turning to me and emphatically declaring, "Jeff, these animals *really* exist! What do you plan to do about it?"

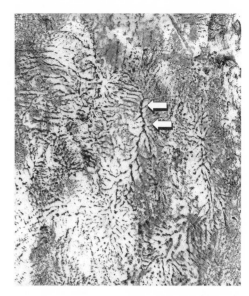

Close-up of ridge detail showing healed scar

Some months after Chilcutt's visit, I received a purported sasquatch footprint cast to examine from Steve Hyde. The cast had been made in 1997, by an on-duty deputy sheriff James P. Akin, while responding to a disturbance at a ranch on Elkins Creek, in Pike's County, Georgia. Tracks were found in the creek both underwater and on an exposed silt bar. There were areas of noticeable skin ridge detail indicated at several

Cast of a 13-inch footprint from Onion Mountain Road, northern California, 1967, showing coarse ridge detail

Taped lift of the skin ridge detail of the cast from Blue Creek Mountain Road

points on the cast. Officer Chilcutt was consulted, and after a thorough examination of the cast, determined, ". . . the dermal ridges in areas 'A' and 'B' [the stems of the second and fourth toes] are definitely the dermal ridges of a nonhuman primate. This conclusion is based on the fact that humans have creases running perpendicular to the lateral ridges on the first joint of the toes, where the toe meets the foot. No such creases were observed in areas 'A' and 'B.' In area 'C' [medial side of the foot near the base of the hallucial metatarsal] the ridges flow lengthwise along the side of the foot. This does not occur in the human or the known nonhuman primate. This ridge flow is also consistent with the 1967 Blue Creek Mountain Road cast and the 1984 Walla Walla, Table Springs cast."

As with Krantz before him, Chilcutt's conclusions have been discussed with several of his associates in law enforcement who have been very intrigued by the evidence. His preliminary findings have received typically guarded reactions from anthropologists and dermatoglyphists, even being denied an airing at some professional conferences. Certainly

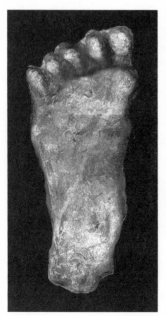

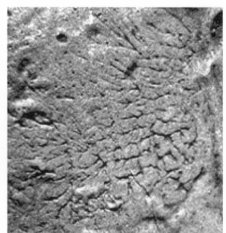

18-inch cast from Elkins Creek, southwestern Georgia; close-ups of two areas showing coarse skin ridge detail (Courtesy of Jimmy Chilcutt)

they warrant further academic review. If repeated independent occurrences of dermatoglyphics in sasquatch footprint casts spanning several decades, with hundreds of miles of geographic separation, and displaying consistent yet distinct features of ridge texture and details of flow pattern can be confirmed, it would constitute compelling evidence for an unknown primate.

15

SPLITTING HAIRS AND MOLECULES: DNA AND PHYSICAL EVIDENCE

Footprints, with all they imply, constitute no more than trace evidence, and no matter how compelling they may be, no matter how anatomically appropriate and internally consistent they appear, they are merely suggestive of the presence of the sasquatch. Where is the *hard* evidence? The vexing absence of bones and teeth has already been discussed in light of the nature of soil chemistry in Northwestern forests. Hair provides another *physical* remnant of the creature itself. Precise hair identification—trichinology—is, however, a challenging process requiring an extensive reference collection with which to compare the unknown sample. There can be considerable individual variation in hair length, color, texture, and stage of growth within a species, as well as variation between different regions on the body of a single individual animal. Arriving at a conclusive identification can be a labor-intensive process requiring exhaustive comparisons with known standards.

In the past there have been several independent analyses of hair attributed to sasquatch, more often than not conducted at the request of an amateur investigator. Usually, the hairs were readily identified as belonging to a commonly known animal such as bear, coyote, or human. However, some hair samples were of indeterminate identity. While an indeterminate indentification of an alleged "sasquatch hair" is interpreted by some as indication of an unknown animal, it is more conservatively regarded by others as the lack of a comprehensive collection of hair samples from known species of animals with which to compare the strand in question. Indeed this would be the only reasonable outcome for hair that might in fact have come from a sasquatch. All that could be concluded is what species the hair did not appear to belong to. Conclusive identification depends on a match to a known sample of hair, i.e., an established standard. Without a confirmed sample of sasquatch hair, any hair truly originating from a sasquatch would necessarily languish in the indeterminate category. Such a standard is unlikely to be acknowledged until hair is pulled directly from a sasquatch body by a qualified analyst.

Hairs obtained from the Bluff Creek area by Ivan Sanderson were submitted for analysis to Dr. F. Martin Duncan, who was in charge of the extensive hair collection at the London Zoo. Duncan stated that they could not be matched to any known North American mammal, but that they could be from an unknown large primate, based on common features shared with species of that order.

Nearly a decade later, suspect hairs collected in central Idaho, in 1968, were sent to Ray Pinker, an instructor of police science at California State College in Los Angeles. He likewise determined that the specimens did not match any sample of known animal that he had in his collection, but showed some characteristics common to humans and nonhuman primates. The hairs were light, showing variation in color and thickness along their length unlike human hair, but exhibited a humanlike scale pattern and lacked a continuous medullary core. The sample contained both coarse guard hairs and fine underhairs, uncharacteristic of primates. As suggestive as these observations might appear, Pinker reiterated that he could not identify the hairs as sasquatch until he had a sample of authentic sasquatch hair to match it to.

Dr. Sterling Bunnell, M.D., a member of the California Academy of Sciences, conducted another hair analysis in 1993. Of a sample collected in northern California, he said, "I have examined the hair specimen you provided from Damnation Creek and compared it by light microscopy under direct and transmitted illumination with human, chimpanzee, gorilla, orangutan, and *Pygathrix* monkey hair. It is clearly related to the human-chimpanzee-gorilla group, but is distinguishable from each of these. On the basis of surface and internal structure it seems more like gorilla hair than human or chimp, while by the same criteria human and chimp appear closely related. The specimen hairs are remarkable in the extremely fine and diffuse pigmentation (the other species show dark melanin clumps and medullary streaks) and the absence of observable medullary structure."

Such results have remained unimpressive, or simply unknown, to the scientific community at large. But as I discussed this challenging situation of the lack of a sasquatch standard with Dr. Henner Fahrenbach, it occurred to us that if the indeterminate hair samples did in fact come from an unknown species of ape, there should be a suite of distinctive and distinguishing diagnostic features consistently characteristic of the suspect hairs. By backtracking to some of the labs and collecting their reports, it did appear that the indeterminate hairs displayed notably uniform characteristics, and likely represented an uncataloged single species of animal as their source. What are these characteristics of hair that form the basis of identification?

In nonprimate mammals, the hair is differentiated into three types: 1) longer, coarser guard hairs, which provide mechanical protection, 2) a finer undercoat of fur hairs, which provide insulation, and 3) stiff vibrissae or whiskers, which serve as tactile organs. Of these types, the guard hairs are of greatest diagnostic value. Primate hair is generally undifferentiated and has been described as possessing a modified combination of the characteristics of guard hairs and undercoat hairs. All hair varies, even single hairs from the same individual. In this way, hairs are rather like snowflakes. Therefore, a composite of characteristics occurring in all *or most* of the hairs in the reference sample is considered diagnostic. The outer layer of the hair shaft is called the cuticle. It consists of overlapping scales and may be thick or thin, clear or pigmented. The intermediate layer is the cortex, the main constituent of the shaft that gives the hair its texture. It contains keratinized cells and one of two types of pigment granules: melanin (brown) and phaeomelanin (blond and red). Also present are cortical fusi (tiny air bubbles) and ovoid bodies (dark ovals) the functions of which are unknown. An innermost central channel, found only in large thick hairs, is the medulla. It varies in diameter and may be dark when empty and filled with air, or it may be clear, although actually filled with cells.

Other characteristics include the diameter, cross-sectional shape, and length of the hair shaft. The hairs of individual species tend to wear differently. Some hair ends fray and split; others wear to a blunted end. Human head hair is ever-growing, lengthening about 10 cm per year, and typically has a cut end that is easily recognized under magnification. The follicle structure is less well studied, but may also display distinctive features. These are complicated by the changes that accompany the hair's growth cycle. Growing hair is said to be in the anagen phase, and the root appears as a stretched out tangle. During the resting phase, or telogen phase, the root ends in a neat bulb.

Fahrenbach has assembled over a dozen samples of hair that could not be attributed to any of the commonly known animals of North America, but which share a common suite of distinguishing features of the sort just discussed. Two of these samples are described in greater detail here:

Unknown sample A averaged 8" (20 cm) in length (ranging from 3" to 15"), was dark reddish brown to black; wavy to slightly curly in parts; mean diameter of 65 μm (range 58–73 μm); cross-section round to slightly flattened entire length. Medulla absent, minimal fragmentary medulla in one region; cuticle pigmentation showed an irregularly waved mosaic pattern; scale margins were smooth to mildly crenulated (wrinkled); distance between scales intermediate; 9–12 scales per 100 μm linear distance. Most hairs were in the telogen stage of hair growth cycle, that is, they were ready to fall out on their own. A few strands had follicles attached. Some of the hairs had split ends, i.e., a brushlike tip, whereas most had smoothly rounded worn tips. None were cut.

Unknown sample B averaged 3.4" (8.5 cm) in length (ranging from 2" to 5"), was reddish brown to blond. It was identical in anatomy to sample A, except for the complete absence of a medulla. All hairs had attached follicles. Neither sample shows any indication of being cut.

"These samples were taken from adjacent fresh twist-offs," Fahrenbach noted. "They came clearly from the same species, but from two different individuals, one light reddish brown with an average length of four inches (worn ends), the other with dark reddish brown (black to the naked eye), average length eight inches. Two sasquatches matching these colors were observed within twenty minutes in the immediate vicinity of the twist-offs by several eyewitnesses. None of the witnesses had the prerequisite hair color or morphology."

The anatomy of the hair samples ruled out the common mammals of the region. There was no feature that absolutely ruled out identification as human hair, although the lack of cut ends on relatively short hairs seemed to make a human source quite unlikely. "The glitch in the absolute determination of the hair," conceded Fahrenbach, "is the fact that *some* human hair can have a similar appearance." And yet the circumstantial evidence of the discovery combined with the anatomy seemed to leave the sasquatch option open.

As novel molecular techniques became available, they held the potential of a surer identification of the hair than sole reliance on anatomy. In 1988, Bob Titmus sent some hair samples collected in northern California to Dr. Jerold Lowenstein, M.D., at the University of California, San Francisco. Dr. Lowenstein had pioneered techniques of radioimmunoassay. This technique employs the body's ability to generate antibodies to foreign antigens. Antibodies were made to the unknown hair sample and then were cross-reacted with human, gorilla, and chimpanzee samples. Lowenstein concluded, "I used all of your material to make antibodies in a rabbit and then tested that antibody to see which animal's serum gave the best reaction. It reacted more or less equally with human, chimpanzee, and gorilla, which suggested that it was probably human. Those three are so close that I can't distinguish them (yet) [sic] on the basis of hair, though I can distinguish them if I have skin or blood." Lowenstein's conservative conclusion may not have taken into account the fact that the hairs in question were grown to length, with uncut ends, making it very unlikely that they were human.

The recent advent of polymerase chain reaction (PCR), a process that permits the exponential amplification of a targeted strand of DNA, and the ensuing revolution in DNA research has opened a new possibility for the study of sasquatch hair. DNA can be extracted from the cells within the hair and selected portions of the molecule can be sequenced. The sequence of the nucleotides, or DNA subunits, can in turn be compared to homologous sequences for known species and the sample may be identified. If it does not match the sequence of any known animal, then at least the phylogenetic position of the sample is determined. In other words it can be determined what its closest relatives are in its respective family tree. As with the hair itself, without a known sample of sasquatch DNA, the DNA sequence cannot be conclusively attributed to sasquatch.

Fahrenbach's samples A and B, described above, were examined by Dr. Paul Fuerst, at Ohio State University. Repeated attempts using both shafts and follicles, failed to successfully amplify mitochondrial DNA from the samples. Several reasons were hypothesized to explain the obstacle: the small diameter of the hair combined with the lack of a cellular medulla, the duration and conditions of storage of the hair before analysis, or some species-specific property of the DNA.

Todd Disotell of New York University has candidly said, "It's definitely true that different species' hair shows differential amplification probabilities. For instance, chimps' work well, gorillas' are a challenge. The closer to the bulb or the bulb itself is clearly the best. Shafts don't always have DNA in them. Unfortunately, I know of no study that has studied it intensively or in a controlled manner that could pinpoint what works best. Basically, you just try and cross your fingers."

Curiously, one quest for yeti DNA had a similar outcome. An expedition to Bhutan retrieved hair samples of interest, collected from an alleged yeti nest. These were conveyed to the Oxford molecular biologist, Bryan Sykes, for analysis. Sykes began with the assumption that he was dealing with bear hair, and so his findings came rather as a surprise. Attempts to sequence the DNA were altogether unsuccessful. In an interview with *Smithsonian Magazine* he remarked, "We normally wouldn't have any difficulty at all. It had all the hallmarks of good material. It's not a human; it's not a bear, nor anything else that we have so far been able to identify. We've never encountered any DNA that we couldn't sequence before. But then, we weren't looking for the yeti . . . I didn't think this would end in a mystery."

Not all yeti samples have proved so mysterious. Peter Matthiessen submitted a hair sample from Tibet for analysis. Despite the insistance of his sherpa guides that the hair was from a yeti, the analysis concluded that the coarse hair was from a horse. However this does little to rationalize the indeterminate nature of the Bhutan hair studied by Sykes, or the unsequenced samples of alleged sasquatch hair. It should not be taken as affirmative evidence of either reputed creature, but it certainly signifies and should be acknowledged that an unanswered question remains that demands scientific attention.

When the Skookum cast was recovered, Noll, Randles, and Fish left the encrusting soil in place until I was able to arrive and direct the cleaning process. Using optivisors and magnifying glasses, we meticulously cleaned away the dirt using dental picks, needle probes, and soft brushes. Every fiber that was encountered was carefully placed in paper specimen envelopes. Dr. Fish and I subsequently sorted these. It was obvious that the hair of a variety of animals was present at the site, as was to be expected given the presence of various tracks. However, some hair appeared distinct during our preliminary assessment, and samples of those hairs of particular interest were sent to Dr. Fahrenbach for independent examination. He concurred that some of the hairs were consistent with the profile for sasquatch hair that he had established. Subsequently Dr. Fish conducted his own microscopic evaluation and identified additional specimens that coincided with the sasquatch characteristics.

The recovery of suspect hairs from the Skookum cast provided a new opportunity to reattempt DNA analysis. For this endeavor, Dr. Craig Newton, a molecular biologist with British Columbia Research, was enlisted. Newton indicated that, "If we get a DNA sample, we will generate a DNA sequence that will identify the source of the hair somewhere on the tree of life, either between humans and primates, or between other organisms of the forest. It's all in the DNA."

Molecular biologist Craig Newton attempted to sequence DNA from sasquatch hair samples. (Courtesy of Doug Hajicek / Whitewolf Entertainment, Inc.)

Newton's intention was to get as much data as possible in order to attract the attention of experts in *primate* DNA and human forensics to examine the question more thoroughly. He successfully extracted DNA from the hair samples, although it was fragmentary. In this case he attempted to utilize nuclear DNA primers, the short bits of DNA used to initiate the amplification of the target DNA. Preferable would have been primers for mitochondrial DNA (mtDNA). The mitochondria are numerous organelles within the cell that produce chemical energy and have their own strand of DNA. The mtDNA has characteristics making it sometimes more useful for amplification, identification, and analysis. Based on the sequences yielded by the fragmentary nuclear DNA, he was unable to rule out human contamination, or even a possible human source.

The study of hair samples has been troubled by the occurrence of synthetic fibers mistakenly associated with sasquatch activity, or perhaps intentionally passed off as sasquatch hair for whatever reason. Paul Freeman displayed examples of sapling trees twisted over, presumably by sasquatch. Entangled in the splintered wood were what appeared to be hair strands. These turned out to be extruded monofilament Dynel fibers, commonly used as synthetic fur and wigs, as well as stuffing in furniture. On the surface, this appears to be a clear-cut case of hoaxing. However, others, including a retired game warden, have also discovered suspicious "hair" that likewise turned out to be similar synthetic fibers. It has been suggested that these resilient fibers have become something of a pervasive environmental contaminant, although the extent of this has not been determined. It should be noted that Freeman has collected several samples of true hair that number among Fahrenbach's collection of possible sasquatch hair, including samples from which degraded DNA was extracted by researchers at Ohio State University. It seems unjustified to throw out all the evidence as the result of a case of misidentification.

Scat, the excrement or feces of an animal, constitutes another important form of physical evidence. It is a significant form of sign that conveys revealing information about the identity of the animal, the location of its activities, and its size and diet. In the past the study of scat (scatology) has been largely descriptive. Identification of scat is based on characteristics such as texture, shape, size and quantity, color, odor, and location. There is a great deal of inherent variability in the appearance of scat. For example, tracker Jim Halfpenny observes, "Experience has shown that visual identification of scat without additional clues may be correct only 50–66 percent of the time. Experienced naturalists in laboratory tests of unmarked scat correctly identified only 88 percent of lion scat and 64 percent of bobcat scat. The wrong identifications stress the need for other supporting clues, including tracks, wherever possible."

Since bears are large omnivores, it stands to reason that their scat, based on the effects of a similar diet, would most likely be erroneously attributed to sasquatch. Therefore, a detailed description of the characteristic and seasonal features of the feces of the bear family follows, as provided by Jim Halfpenny in *A Field Guide to Mammal Tracking in North America* (Johnson Books, Boulder, CO). "Bear scats are thick cords with blunt ends. The quantity of scat is often great, and you may see large piles. The omnivorous bears have brown, black, and blue-colored scat from eating pine nuts, animal protein, and berries respectively. Bear scat often contains insects, especially ants, termites, and bees. You may have to look hard at the scat to identify the undigested

Sample of scat tentatively attributed to sasquatch (Courtesy of John Green)

heads of these insects. Plant remains are also common, with grasses, dandelions, horsetails, and thistles being favored. Some scat containing plant material will be black. The fibrous nature of the scat and sweeter plant smell will separate these scat from those containing animal protein. In the Fall, a diet of acorns produces above-average–size brown scat. From April to October, grizzlies will raid the caches of squirrels for white-bark pine nuts and the husks will be evident in the scat. A Fall diet of berries will produce soft to semi-liquid scat. Scats greater than 2 inches (5 cm) in diameter have been considered grizzly scat. However, Herrero reported that of 140 grizzly scats measured in Banff National Park, 60 (58 percent) were less than 2 inches. Larger grizzlies produce larger scat. If you use the 2-inch rule, you will tend to misidentify the females, who may be with their young, and young males, a very serious error."

Innovations in the study of endoparasites, chemistry, and DNA can potentially provide added information. Ivan Sanderson reported the results of a rare analysis of endoparasites conducted by an unnamed Oregon medical lab on a scat sample suspected to have originated from a sasquatch. It contained the eggs of a parasitic nematode worm of the genus *Trichuris*. The size and proportion of the eggs are distinct for individual species, which in turn are very host-specific. That is, certain parasite species are found only in the guts of particular host species of mammals. The sasquatch sample contained three types of eggs, although the examiner demurred from identifying them due to their state of deterioration. It was concluded that the largest eggs were outside the range of human parasite ova, but that such large eggs had been reported from various other primates. Sanderson also reported on a sample that was packaged in a container with dry ice and sent to him in New York for transfer to Dr. W.C. Osman Hill, then senior scientist at the London Zoological Society. In Sanderson's words, "This specimen shook up the scientists . . . this fecal mass did not in any way resemble that of a known North American animal. On the other hand it did look humanoid, but it had some peculiar features, as if the lower bowel had a spiral twist. But, above all, it was composed exclusively of vegetable matter, and this, as far as could be identified, of local (Californian) fresh-water plants. The real clincher was that it contained the eggs and desiccated remains of certain larvae otherwise only known in (a) some North American Indian tribal groups in the Northwest, (b) pigs imported from south China, (c) human beings in country districts in southwest China, and (d) pigs from that same area." Another potential Asian connection?

Dr. Vaughn Bryant, an anthropologist at Texas A&M University, examined two scat samples from the Pacific Northwest, which consisted largely of plant matter. Based on the microscopic examination of the reconstituted cells of the constituents of the feces, he was able to rule out man, moose, elk, deer, and bear as the source in these cases, but could not identify the source further. His findings were published in the edited collection of papers, *Manlike Monsters on Trial*. Bryant reflected on the course of events: "I lived in the Pacific Northwest during the late 1960s and 1970s when I taught anthropology at Washington State University. I was good friends with Grover Krantz, and he and I would discuss much of the ongoing search and information about sasquatch. I was a skeptic, but open-minded. We went out to look at sighting locations once in a while. When I moved to Texas, I continued my interest in the subject and let it be known that I was willing to examine feces and hair if any were found (these are my specialty areas). Soon, I was receiving all sorts of packages of feces—some still fairly fresh. The problem was that I was soon overwhelmed with samples and could not examine all of the samples due to time and expense. Remember, I was also teaching a full load, working with my own graduate students, trying to publish research, establish a scientific reputation, and do what my university and colleagues called 'scholarly work.' Bigfoot did not fit into that category and, in fact, I was told several times that my continued interest in the subject put my career in jeopardy. Nevertheless, I continued working on some of the materials. The bottom line is that I could never find anything that would eliminate the fecal samples from belonging to some other animal that was also omnivorous. Perhaps someday someone can do DNA of a sample. However, the process is costly and time consuming."

Clearly the question of sasquatch's existence is unlikely to be resolved conclusively without physical evidence. The conventions of zoological taxonomy require a type specimen, traditionally in the form of a body or a sufficiently diagnostic physical body part, to decisively establish the existence of a new species. Whether DNA alone will ultimately satisfy that standard remains to be seen. I am doubtful. I am not aware of a precedent for determining a new species on the basis of DNA evidence, in the absence of a physical specimen. Subpopulations of recognized species have been differentiated based on DNA sequence differences and been given species status, but those subpopulations were previously known and sampled. However, a substantial DNA sequence would be a major development in the search. Short of a body, or a significant part of one, it would seem that the elusive DNA evidence is the next objective to be focused upon by those who can no longer dismiss the matter out of hand.

Where We Stand: The Evidence Weighed and Measured

The most stifling situation that can ensue from this subject is for the treatment of the question of the existence and nature of sasquatch to be reduced to an argument between "believers" and "skeptics." I am frequently asked, "Do you believe in sasquatch?" I invariably and firmly reply that a question of *belief* is simply not at issue. Belief generally connotes the acceptance of something as true in the absence of objective evidence or conclusive proof. It is usually equated with a position of faith. Science is about subjecting hypotheses to evaluation by marshaling evidence that may either refute, or lend support to a premise. Therefore, from a scientific standpoint I can say that a respectable portion of the evidence I have examined suggests, in an independent yet highly correlated manner, the existence of an unrecognized ape, known as sasquatch. This conclusion of necessity remains tentative and provisional since the interpretation of the evidence, however persuasive it may be at this point, remains ultimately inconclusive. It should be well noted that a pending conclusion has rarely provoked a scientist to abandon research that is backed by empirical evidence. Unfortunately, those scientists who *have* made the effort to systematically review and evaluate the data, and as a result are motivated, if not obliged, to pursue their intellectual curiosity further, are frequently labeled as "believers" and are judged incapable of further objectivity.

From the sidelines, ideological skeptics would wrap themselves in the banner of science and profess that they approach a controversial phenomenon from a rational and critical position. However, the extreme conservatism typically embraced by such individuals is not without its own pitfalls. Michael Shermer, founder of the Skeptics Society and publisher of the *Skeptics Magazine,* offers this caution in his manifesto: "The key to skepticism is to continuously and vigorously apply the methods of science to navigate the treacherous straits between 'know nothing' skepticism and 'anything goes' credulity." This is a worthy aim, and Shermer's caution about the treacherous nature of the passage is warranted, the correct bearing to that middle course is sometimes as elusive as sasquatch itself. It would seem that many skeptics run aground as they unwittingly list or deliberately steer toward the extreme of incredulity. This can result from a lack of motivation to become informed about the essential evidence, the simple impracticality of thoroughly evaluating every assertion that comes along, or even more fundamentally, the sheer lack of the requisite expertise to evaluate the evidence for such a broad range of purported phenomena. Whichever cause it may be, many professing skeptics do not allow their own ignorance of the primary data to prevent them from pronouncing baseless and often cynical condemnation of the subject and those most familiar with it.

In his regular *Scientific American* column, Shermer was apparently prompted by the rash of publicity surrounding the Wallace story to address the issue of Bigfoot. In opening the column he professed open-mindedness, saying that Bigfoot, ". . . while occasionally eliciting an acerbic snicker, enjoys greater plausibility [than the Jackalope] for a simple evolutionary reason: large hirsute apes currently roam the forests of Africa, and at least one species of a giant ape—*Gigantopithecus*—flourished some hundreds of thousands of years ago alongside our ancestors." Then Shermer runs aground upon a shoal on the coast of "know nothing" skepticism. He asserts that although the question of Bigfoot may have warranted our exploratory resources at one time, it no longer does. Why? Because he believes that "If such creatures survived in the hinterlands of North America and Asia, surely by now one would have turned up." After rehearsing this tired cliché, he next fully exposes his lack of familiarity with the details and depth of the evidence by summing up the entire matter as simply a collection of stories. According to science historian Frank J. Sulloway, of the University of California at Berkeley, stories do not constitute science, and it makes no difference whether one confronts one, ten, or a hundred anecdotes. Shermer raises this caveat to a maxim when dealing with "Bigfoot hunters." He admonished, "Their tales make for gripping narratives, but they do not make sound science. A century has been spent searching for these chimerical creatures. Until a body is produced, skepticism is the appropriate response."

Indeed, in so far as skepticism embodies the *continuous and vigorous application of the methods of science,* then Shermer's conclusion is quite justified. However, his characterization of the question does little to convey a sense of its history or of the mounting evidence—trace, photographic, and physical—such as discussed in this volume, which takes the matter far beyond a mere collection of anecdotes. Shermer was recently invited to present a public lecture on the Idaho State University campus. He didn't on this occasion broach the subject of sasquatch, but he was engaged during his visit by a group of students who happened to be producing a documentary called, "Rocky Mountain Bigfoot," for their mass communications class project, mentored by Dr. Michael Trinklein. During a brief recorded interview, Shermer acknowledged the possibility of the existence of sasquatch, but deemed it very unlikely. "Okay, fine, I'm glad you're looking. I hope you find something—call me when you do. After a hundred years I'd get a little discouraged by now. I'd go try to find something else—like aliens," he said. The students quickly realized that Shermer's skepticism was not based on a familiarity with, let alone expert evaluation of the evidence, and in the face of their probing questions he simply fell back to the position that "It's just mythology. . . . For this to be a science you've got to have a body. That's the way it works."

I should reiterate my acknowledgment that the conventions of zoological taxonomy require a type specimen to establish the existence of a new species. On this point Shermer is in principle quite correct. That having been said, the position that the evaluation of the evidence presented in this book falls outside the purview of good science is frankly indefensible. I don't think any reasonable investigator is out to convince the skeptics in the absence of physical proof, but I suggest that to offhandedly ignore or simply dismiss the current "body" of evidence that bears on this question, is to have failed to navigate the treacherous strait between "know nothing" skepticism and "anything goes" credulity. The nature and extent of the evidence fully justifies, in truth demands, the serious attentions of scientists.

Perhaps even more disingenuous is to acknowledge the evidence, but then to distort and make light of it. Benjamin Radford, managing editor of *Skeptical Inquirer* magazine, published a skeptical summary entitled, "Bigfoot at 50: Evaluating a Half Century of Bigfoot Evidence." Assuming that the Bigfoot phenomenon abruptly sprang into existence in 1959, Radford acknowledges that after nearly fifty years, the question of Bigfoot's existence remains open. He notes that the issue comes down to evidence. "There is indeed no shortage of evidence. The important criterion, however, is not the *quantity* of the evidence, but the *quality* of it." To support his contention that the evidence for Bigfoot is generally poor, he cites contradictory conclusions drawn by "Bigfoot researchers," but makes no distinction between the conclusions drawn by credentialed scientists and those of amateur investigators. Instead, he pits their interpretations of the data against one another as if they were on equal footing. As each category of evidence is necessarily summarized briefly, but incredibly superficially, he repeatedly misleads, misrepresents, and selectively quotes the opinions of unqualified individuals as if they were authorities. Eyewitness accounts are summarily dismissed, even those by trained observers, such as wildlife biologists, since, as he puts it, "Anyone, degreed or not, can be mistaken." All spurious footprints are lumped together and then he wonders at the lack of consistency that one might expect from a single unknown species. Turning to the Skookum imprint, instead of evaluating the features of the cast itself, he remains incredulous because the animal "made such an odd approach to the food." Not so odd if he had a familiarity with great ape feeding behavior.

Radford's discussion of the Patterson-Gimlin film is most egregious—rife with inaccuracies that are implied as indicative of a probable hoax. He states the assumption that Patterson was on the scene with the express purpose of capturing a sasquatch on film. In point of fact, Patterson was there in response to the recent discovery of clear footprints with the intent of filming fresh tracks for inclusion in a documentary he intended to produce. Next, Radford says that a known hoaxer (Ray Wallace) claimed to have told Roger exactly where to go to see the Bigfoot on that particular day. Indeed, Wallace *claimed* to have told Roger where to go, but it was clear from later interviews that he possessed little knowledge of the specific area, and his family, although claiming he had hoaxed other films and photos, publicly denied any involvement by Wallace in the Patterson incident (although in the face of waning media interest in their story, some members of the family have made public statements to suggest or even assert that Wallace was responsible for the Patterson-Gimlin film as well). Finally, Radford alleges that Patterson turned quite a profit and garnered publicity for his book. But, in fact, what modest profits may have been realized were expended in further efforts to pursue the creature. Even when Patterson was about to suffer a premature death from Hodgkin's disease and could have reaped significant profits by revealing a fake, he maintained his original account of the events, as has his disenfranchised partner in the filming, Bob Gimlin, to this day.

At last, Radford equates *inconclusive* results of the analysis of hair, scat, and blood samples that defy identification with negative evidence (i.e., the lack of information) and glibly suggests that when a *conclusive* result is found, the source is commonly elk, bear, or cow hair. This approach ignores the fact that a wide sample of independently collected hair defies identification or standard DNA sequencing protocols. That conclusion is quite different from negative evidence.

Both of these skeptics are biased by a different but related misconception. Shermer asserts that a century has been spent searching for Bigfoot with no conclusive result to show for it. *Who* has been doing the searching? What institutionally sponsored and equipped scientific expeditions have committed sustained effort and resources to document or alternately refute the existence of a North American ape? Virtually none. On the other hand, Radford implies that the evidence for sasquatch has been weighed and measured and has been found wanting. Radford and the majority of those critics cited in his article have limited expertise to evaluate the diverse evidence—e.g., footprints, dermatoglyphics, hair, scat, vocalizations, etc.—with a degree of competence or authority. Indeed, precious few qualified scientific researchers have made any serious attempt to systematically review and evaluate the data, although the number continues to increase. Several such scientists have been featured in this volume, and their approach to the subject and the conclusions they arrived at contrast markedly with those of the professional skeptics—and of some of the "believers" for that matter.

A welcome sign of a thawing of attitude among the "orthodoxy" came as a sidebar in a *Scientific American* article. The article in the August 2003 issue was by paleoprimatologist Dr. David Begun, of the University of Toronto. The sidebar was entitled, "Bigfoot Ballyhoo." The text was reasonably and refreshingly objective in addressing the question, but the editorial choice of title betrays a persistent underlying "tongue-in-cheek" attitude and an underappreciation of the accumulating evidence suggesting the potential existence of a North American ape. *Ballyhoo* is defined as an "exaggerated clamorous attempt to advance a cause." Implicit is the suggestion that the "believers" are stumping for sasquatch by exaggerating the weight of the evidence. The more disturbing exaggeration I have witnessed is the excessive trivialization of the data, often by the media, but also by self-assured skeptics and by inadequately informed scientists.

What are the implications of indeterminate hair samples that display consistently distinct characteristics; many hundreds of eyewitness encounters by reputable professionals, including wildlife biologists, foresters, national park rangers, field geologists, and law enforcement officers; functional analyses of an extensive collection of footprints, some that I have personally documented in remote areas of northern California and Washington; dermatoglyphics; recorded vocalizations; and films that have withstood the critical scrutiny of locomotion experts? The more notable clamor seems to arise from those who emphatically refuse to examine these data or to discuss them in an objective and scientific manner.

Rather than consider the sasquatch some uniquely extraordinary apparition, it is much more productive to identify those features it appears to bear in common with other organisms. In what ways is it consistent with our knowledge of the evolution, natural history, and biogeography of primates on the one hand and with known wildlife in North America on the other? In this book the mounting evidence for the existence of an unknown North American primate has been considered by a world-class lineup of specialists from a broad spectrum of disciplines. They have ventured to weigh and measure the evidence concerning the presence of such a giant North American ape on its own merits without preconception or prejudice. What conclusions have they drawn from their investigations? Have their previous opinions been modified by the experience?

In several instances the evidence is intriguing if not outright persuasive and compels to further investigation. The footprints and Skookum cast were pivotal in turning the heads of noted primate anatomists and dermatoglyphists. Expert viewers could not casually dismiss a number of films and videos as simply an obvious case of a "man in a fur suit." Morphologically, the hair evidence indicates some unknown animal, although the definitive DNA evidence remains elusive and must become a focus of future endeavor. The pervasive sightings and notable correlations of consistent anatomy, behavior, and vocalization also point to a real animal—one that displays remarkable parallels to known great apes, while exhibiting unique characteristics appropriate to its particular ecology. The deeply rooted cultural and historical knowledge of a hairy wildman possessed by indigenous populations attests to a zoological entity.

Where does the investigation go from here and what data are required to reach a definitive resolution? A body? DNA? The final answers to these questions will require a challenge to some preconceptions held by the scientific community, in which extreme skepticism is sometimes deemed a requirement for reputable membership. For me, it now seems more incredible to suggest this matter could all be dismissed as mere stories, misidentifications, and spurious hoaxes than it is to at least rationally entertain the well-founded suggestion that the legend of sasquatch possibly has its basis in a real animal and may eventually prove to be among the most astounding zoological discoveries ever.

What might be the implications of such a discovery? The thing that clearly unites us *physically* as a global human family is our common origin. Our understanding of that common physical origin has been increased as we have come to better understand and appreciate our closest kin in the animal kingdom—the great apes. The pioneering researches of early field primatologists have virtually pulled back a curtain, revealing the narrowness of the "gulf" that separates us from them. How much *more* impacting would be the discovery and recognition of a hominoid possibly more humanlike than any other inhabiting this planet? As the shrinking tropical habitats of the known apes have brought to public consciousness the stewardship we bear for the distant tropical environment and its imperiled coinhabitants, so might the recognition of sasquatch sharpen our sense of responsibility for our own temperate habitats and bring into finer resolution our connection to wild places.

Selected Bibliography

Alley, J. R. *Raincoast Sasquatch*. Blaine, Wash.: Hancock Publishers, 2003.

Arcadi, C., D. Robert, and C. Boesch. (1998) Buttress drumming by wild chimpanzees: Temporal patterning, phrase integration into loud calls, and preliminary evidence for individual distinctiveness. *Primates*. 39(4):505–518.

Bayanov, D. *America's Bigfoot: Fact, Not Fiction*. Moscow, Russia: Crypto-Logos Publishers, 1997.

Bayanov, D., I. Bourtsev, and R. Dahinden. (1984) Analysis of the Patterson-Gimlin film: why we find it authentic. In: V. Markotic and G. Krantz (eds.) *The Sasquatch and Other Unknown Hominoids*. Calgary: Western Publishers, pp. 219–234.

Begun, D. Planet of the Apes. *Scientific American*, August 2003, pp. 74–83.

Bindernagel, J. A. *North America's Great Ape: The Sasquatch*. Courtenay, B.C.: Beachcomber Books.

Bourne, G. H., and M. Cohen. *The Gentle Giants: The Gorilla Story*. New York: G.P. Putnam's Sons, 1975.

Brown, P., T. Sutikna, M. J. Morwood, R. P. Soejono, E. Jatmiko, Wayhu Saptomo, and Rokus Awe Due. A new small-bodied hominin from the Late Pleistocene of Flores, Indonesia. *Nature* 431:1055–1061.

Bryant, Jr., V. M., and B. Trevor-Deutsch. Analysis of feces and hair suspected to be of sasquatch origin. In: Halpin, M. and Ames, M. M. (eds.) *Manlike Monsters on Trial: Early Records and Modern Evidence*. Vancouver: University of British Columbia Press, 1980, pp. 291–300.

Burns, J. W. Introducing B.C.'s hairy giants. *MacLean's Magazine*, April 1, 1929, p. 9f.

Chorvinsky, M. (1994) New Bigfoot photo investigation. *Strange Magazine*, No. 13, Spring 1994, pp. 10f.

Chorvinsky, M. (1993) Our strange world: Bigfoot and the Ray Wallace connection. *Fate*, November, pp. 22–29.

Churchill, E., McConville J.T., Laubach, L. L., Churchill, T., Erskine, P. and Downing, K. (1978) *Anthropometric Source Book, Volume II: A Handbook of Anthropometric Data*. NASA Reference Publication 1024 (NTIS No. N79-13711/3/3/XPS). National Aeronautics and Space Administration, Scientific and Technical Information Office.

Ciochon, R., J. Olsen, and J. James. *Other Origins: The Search for the Giant Ape in Human Prehistory*. New York: Bantam Books, 1990.

Coleman, L. *Bigfoot! The True Story of Apes in America*. New York: Paraview Pocket Books, 2003.

Cronin, Jr., E. W. *The Arun: A Natural History of the World's Deepest Valley*. Boston: Houghton Mifflin, 1979.

Daegling, D. J. *Bigfoot Exposed: An Anthropologist Examines America's Enduring Legend*. Walnut Creek, Calif.: AltaMira Press, 2004.

Daegling, D. J. (2002) Cripplefoot hobbled. *Skeptical Inquirer* 26(2):35–38.

Daegling, D. J., and D.O. Schmitt. (1999) Bigfoot's screen test. *Skeptical Inquirer* 23(3):20–25.

Daegling, D. J., and Grine, F. E. (1994) Bamboo feeding, dental microwear, and diet of the Pleistocene ape *Gigantopithecus blacki*. *South Africa Journal of Science* 90:527–532.

Dennett, M. 1989. Evidence for Bigfoot? An investigation of the Mill Creek 'Sasquatch Prints.' *Skeptical Inquirer* 13(3), Spring: 264–272.

Ducros, A., and J. Ducros. From satyr to ape: A scandalous sculpture in Paris. In: R. Corbey and B. Theunissen (eds.) *Ape, Man, Apeman: Changing Views Since 1600*. Leiden: Department of Prehistory, Leiden University, 1995, pp. 337–339.

Enders, R. K. The "Abominable Snowman." *Science* 126:858 (25 October 1957).

Fahrenbach, W. H. (1997–1998) Sasquatch: size, scaling, and statistics. *Cryptozoology* 13: 47–75.

Galdikas, B.M.F. *Reflections of Eden: My Years with the Orangutans of Borneo*. Boston: Little, Brown and Co., 1995.

Gee, H. *H. flores,* God and Cryptozoology. *www.nature.com/news/2004/041025/full/041025-2html*

Gill, G. W. Population clines of the North American Sasquatch as evidenced by track lengths and estimated stature. In: Haplin, M. and Ames, M. M. (eds.) *Manlike Monsters on Trial: Early Records and Modern Evidence*. Vancouver, University of British Columbia Press, 1980, pp. 265–273.

Glickman, J. (1998) Toward a resolution of the Bigfoot phenomenon. Hood River, Oreg.: North American Science Institute (unpublished report).

Goodall, J. *The Chimpanzees of Gombe: Patterns of Behavior*. Cambridge, Mass.: Belknap Press, 1986.

Green, J. *Sasquatch: The Apes Among Us*. Seattle: Hancock House Publishers, 1978.

Green, J. *Year of the Sasquatch*. Agassiz, B.C.: Cheam Publishing, 1970.

Green, J. *On the Track of the Sasquatch*. Agassiz, B.C.: Cheam Publishing, 1968.

Greenwell, J. R. Cryptozoology: On the trial of unverified animals. *Wild Tracks* (Newsletter of the International Wildlife Museum) Winter, 2001:4–5.

Greenwell, J. R., D. J. Meldrum, M. T. Slack, and D. A. Greenwell. (1999) A Sasquatch field project in northern California: Report of the 1997 Six Rivers National Forest Expedition. *Cryptozoology* 13:76–87.

Halfpenny, J. *A Field Guide to Mammal Tracking in North America*. Boulder: Johnson Books, 1986.

Hillary, E. *Nothing Venture, Nothing Win*. New York: Coward, McCann & Geoghegan, Inc., 1975.

Heuvelmans, B. *On the Track of Unknown Animals*. Cambridge, Mass.: MIT Press, 1965.

Husbands, C. When Did 'Hobbit' Humans Die Out? Not So Long Ago, Say Indonesian Villagers. *Berkeley Daily Planet*, November 2, 2004.

Kerbis Peterhans, J., Wrangham, R. W., Carter, M. L., and Hauser, M. D. (1994) A contribution to tropical rain forest taphonomy: Retrieval and documentation of chimpanzee remains from Kibale Forest, Uganda. *Journal of Human Evolution* 25:485–514.

Kirkpatrick, D. (1968) The search for Bigfoot. *National Wildlife* 6(3):43–47.

Kirlin, R. L. and L. Hertel. Estimates of Pitch and Vocal Tract length from Recorded Vocalizations of Purported Bigfoot. In: Haplin, M. and Ames, M. M. (eds.) *Manlike Monsters on Trial: Early Records and Modern Evidence*. Vancouver: University of British Columbia Press, 1980, pp. 274–290.

Kleiner, K. (2000) Bigfoot's buttocks. *New Scientist* 168(2270):8.

Krantz, G. S. *Bigfoot Sasquatch Evidence*. Blaine, Wash.: Hancock House Publishers, 1999.

Krantz, G. S. (1987) A reconstruction of the skull of *Gigantopithecus blacki* and its comparison with a living form. *Cryptozoology* 6:24–39.

Krantz, G. S. (1983) Anatomy and dermatoglyphics of three sasquatch footprints. *Cryptozoology* 2:53–81.

Krantz, G. S. (1972) Anatomy of the Sasquatch foot. *Northwest Anthropological Research Notes* 6(1):91–104.

Krantz, G. S. (1972) Additional notes on Sasquatch foot anatomy. *Northwest Anthropological Research Notes* 6(2):230–241.

Krantz, G. S. (1971) Sasquatch handprints. *Northwest Anthropological Research Notes* 5(2): 145–151.

Leonard, N. L. *The First Americans*. New York: Time-Life Books, 1973.

Long, G. *The Making of Bigfoot: The Inside Story*. Amherst, N.Y.: Prometheus Books, 2004.

Mack, C. *Grizzlies and White Guys: The Stories of Clayton Mack*. Madeira, B.C.: Harbour Publishing, 1996.

Marsh, O. C. (1877) Ancient life in America. *Scientific American Supplement* 4(90):1436–37 and 4(91):1449–51.

Matthiessen, P. and T. Laird. *East of Lo Monthang: In the Land of the Mustang*. Boston: Shambhala Publications, 1995.

McKinnon, J. *In Search of the Red Ape*. New York: Ballantine Books, 1974.

Meldrum, D. J. (2005) Review of *Bigfoot Exposed: An Anthropologist Examines America's Enduring Legend,* by David J. Daegling. *Journal of Scientific Exploration* 19:304–316.

Meldrum, D. J. (2004) Midfoot flexibility, fossil footprints, and Sasquatch steps: New perspectives on the evolution of bipedalism. *Journal of Scientific Exploration* 18:67–79.

Meldrum, D. J. (2004) Fossilized Hawaiian footprints compared to Laetoli hominid footprints. In: D. J. Meldrum and C. E. Hilton (eds.) *From Biped to Strider: The Emergence of Modern Human Walking, Running, and Resource Transport*. New York: Kluwer Academic and Plenum Publishing, pp. 63–84.

Meldrum, D. J. (2002) Midfoot flexibility, footprints and the evolution of hominid bipedalism. *American Journal of Physical Anthropologists Supplement* 34:111–112 (abstract).

Meldrum, D. J. (1999) Evaluation of alleged Sasquatch footprints and inferred functional morphology. *American Journal of Physical Anthropologists Supplement* 28:200 (abstract).

Meldrum, D.J. (1997) Review of *Bigfoot of the Blues,* by Vance Orchard. *Cryptozoology* 13:109–111.

Meldrum, D.J., and D.R. Swindler. (2004) The Skookum imprint: trace evidence of Sasquatch? *Program with Abstracts of the 87th Annual Meeting of the Pacific Division, American Association for the Advancement of Science,* p. 65 (abstract).

Meldrum, D.J., and J.H. Chilcutt (2001) Dermatoglyphics in casts of alleged North American Ape footprints. *Program of the Northwest Anthropological Conference,* p. 35 (abstract).

Meldrum, D.J., and J.R. Greenwell. (1998) Bigfoot: Take two. *BBC Wildlife Magazine* 16:68–71.

Mitani J. C., and J. Stuht. (1998) The evolution of nonhuman primate loud calls: Acoustic adaptation for long-distance transmission. *Primates* 39:171–182.

Montagna, W. (1976) From the director's desk. *Primate News* 14(8):7–9.

Murphy, C.L. *Meet the Sasquatch.* Blaine, Wash.: Hancock House, 2004.

Napier, J. *Bigfoot: The Yeti and Sasquatch in Myth and Reality.* New York: E.P. Dutton & Co., 1973.

Orchard, V. *The Walla Walla Bigfoot.* Prescott, Wash.: Ox-Yoke Press, 2001.

Patterson, R. *Do Abominable Snowmen of America Really Exist?* Yakima, Wash.: Northwest Research Association, 1966.

Patterson, R., and C. Murphy. *The Bigfoot Film Controversy.* Blaine, Wash.: Hancock House Publishers, 2005.

Perez, D. E. Bigfoot at Bluff Creek. BigfooTimes, October 20, 2002 (Released as booklet May 1994).

Pert, C. *Molecules of Emotion: The Science Behind Mind-Body Medicine.* New York: Simon and Schuster, 1999.

Radford, B. (2002) Bigfoot at 50: Evaluating a half century of Bigfoot evidence. *Skeptical Inquirer* 26(2):29–34.

Rennie, J. Unexpected thrills. *Scientific American,* May 1996, p. 4.

Ruby, R.H., and J.A. Brown. *The Spokane Indians: Children of the Sun.* Norman: University of Oklahoma Press, 1970.

Sanderson, I.T. *Abominable Snowmen: Legend Come to Life.* Philadelphia: Chilton Book Co., 1961.

Sanderson, I.T. (1968) First photos of "Bigfoot," California's legendary "Abominable Snowman." *Argosy,* February.

Sanderson, J., and M. Trolle. (2005) Monitoring elusive mammals. *American Scientist* 93: 148–155.

Savage-Rumbaugh, E. S., S. L. Williams, T. Furuichi, and T. Kano. Language perceived: *Paniscus* branches out. In: W. McGrew, L. F. Marchant, and T. Nishida (eds.) *Great Ape Societies*, New York: Cambridge University Press, 1996, pp. 173–184.

Schaller, G. B. *The Mountain Gorilla: Ecology and Behavior.* Chicago: University of Chicago Press, 1976.

Shermer, M. (2003) Show me the body. *Scientific American*, April 14.

Shermer, M. *Why People Believe Weird Things: Pseudoscience, Superstition, and Other Confusions of Our Time.* San Francisco: W.H. Freeman, 1997.

Shipton, E: *The Mount Everest Reconnaissance Expedition.* New York: E. P. Dutton, 1952.

Shuman, J. B. (1968) "This Is Bigfoot." *West, L.A. Times Magazine,* December 15, pp. 24–26, 29.

Simpson, G. G. (1984) Mammals and cryptozoology. *Proceedings of the American Philosophical Society* 128(1):1–19.

Sprague, R., Editorial, *Northwest Anthropological Research Notes* 4(2):127–128.

Sprague, R. Carved stone heads of the Columbia River and Sasquatch. In: Haplin, M. and Ames, M. M. (eds.) *Manlike Monsters on Trial: Early Records and Modern Evidence.* Vancouver: University of British Columbia Press, 1980, pp. 229–234.

Sprague, R. and Krantz, G. S. (eds.) *The Scientist Looks at the Sasquatch II*, Moscow, Idaho: University of Idaho Press, 1979.

Straus, Jr., W. L. Abominable Snowman. *Science* 123:1024–1025 (8 June 1956).

Swan, L. W. "Abominable Snowman." *Science* 126:858 (25 October 1957).

Swan, L. W. *Tales of the Himalaya: Adventures of a Naturalist.* La Crescenta, Calif.: Mountain N' Air Books, 1999.

Swindler, D. R. and C. D. Wood. *An Atlas of Primate Gross Anatomy: Baboon, Chimpanzee, and Man.* Malabar, Fla.: Robert E. Krieger Publishing Company, 1982.

Vlcek, E. (1959) Old literary evidence for the existence of the "Snowman" in Tibet and Mongolia. *Man* 59:133–134.

Wallace, D. R. *The Klamath Knot: Explorations of Myth and Evolution.* San Francisco: Sierra Club Books, 1983.

Ward, M. (1997) Everest 1951: The footprints attributed to the Yeti—myth and reality. *Wilderness Environmental Medicine* 8(1):29–32.

Weidenreich, F. *Apes, Giants and Man*. Chicago: University of Chicago Press, 1946.

White, T. D., and G. Suwa. (1987) Hominid footprints at Laetoli: Facts and interpretations. *American Journal of Physical Anthropology* 72:485–514.

Wylie, K. *Bigfoot: A Personal Inquiry into a Phenomenon*. New York: The Viking Press, 1980.

Yutaka Kunimatsu, Y., B. Ratanasthien, H. Nakaya, H. Saegusa, S. Nagaoka (2004) Earliest Miocene hominoid from Southeast Asia. *American Journal of Physical Anthropology* 124:99–108.

Yerkes, R. M., and A. W. Yerkes. *The Great Apes: A Study of Anthropoid Life*. New Haven: Yale University Press, 1929.

Index

anecdotal evidence, statistical analyses of, 211–19

Aniello, Peter, 218–19

Anthropometric Source Book, 163, 164, 177

Anthropomorpha ("man-shaped" animals), 36, *36*

Antioch (California) *Ledger,* 50

apes, *see* great apes

arboreal apes, 99, 100

Argosy magazine, 151, 161

Athapaskan Indians, 83

Auman, John, 56

Australopithecus afarensis ("Lucy"), 169, 242, 245

axillary organs, 187

B

baboons, 36

bagpipes, 200–201, *201*

Bambenek, Greg, 197, 198, *198*

Barnum and Bailey, 193

Bayanov, Dmitri, 11, *147,* 175

bear, 203–19
  body size, 93
  distinguishing characteristics of, 205
  hibernation, 189
  mistaken sasquatch sightings and, 203, 204, 205, *205,* 207–10
  paws and tracks, 17, 203–204, 206–10, *206–209*
  scat, 267–68
  standing upright, 203, *203,* 204, *205*
  vocalizations, 196
  see also specific types of bear

Beebe, Frank, 167, 171

Begun, David, 91, 275

behavioral parallels of great apes and sasquatch, 179–93

Benson, Robert, *196,* 196–97, 198

Beowulf, epic of, 36

Bergman's rule, 93–94, *94*

Bering land bridge, 95–97

Berry, John, 252

Bhutan's Forestry Department, 41, *42*

Bigfoot
  origins of name, 51, 223
  see also specific aspects of sasquatch

"Bigfoot at 50: Evaluating a Half Century of Bigfoot Evidence," 273–74

Bigfoot Exposed (Daegling), 163

Bigfoot Field Researchers Organization (BFRO), 18, 19–20, 197, 218
  Skookum Meadows expedition, 112–23

Bigfoot of the Blues (Orchard), 23

Big Footprints (Krantz), 23

Bigfoot Sasquatch Evidence (Krantz), 162

Bigfoot: The Yeti and the Sasquatch in Myth and Reality (Napier), *91,* 160–61, 169

Bili (or Bondo) ape, *43,* 43–44

Bindernagel, John, 11, 19, 116, 120, *147,* 181–82, 184, 191, 198–99, 205–206, 222–23, 224

bipedalism, 22, 52, 99–100, 117, 179, 207, 221, 245
  Achilles tendon and, 122
  buttocks and gluteal muscles and, 122
  convergent evolution of, 90, 99, 181, 223, 224
  facultative, 223
  Gigantopithecus and, 99–100, 223
  habitual, 223

Bittner, David, 129–30, *130,* 134–35

G

I

J

K

U

V

W

About the Author

Jeff Meldrum, Ph.D., is an associate professor of anatomy and anthropology at Idaho State University and affiliate curator of the Idaho Museum of Natural History. He received a B.S. and M.S. in zoology from Brigham Young University, specializing in vertebrate locomotion, and a Ph.D. in anatomical sciences, with an emphasis on physical anthropology, from the State University of New York at Stony Brook in 1989. His dissertation research was on terrestrial adaptations in the feet of African monkeys and implications for the evolution of hominid bipedalism. He conducted postdoctoral work at Duke University, where for a time he turned his attention to the evolutionary history of Neotropical primates, participating in paleontological fieldwork in Argentina and Colombia. His first appointment was at Northwestern University Medical School in the Evolutionary Morphology Group. He returned to his home state of Idaho in 1993. At ISU he teaches human anatomy to health-professions graduate students, evolution, and primate studies. His research turns on vertebrate evolutionary morphology, with recurring emphasis on human bipedalism. He has published numerous scientific papers and popular articles. He has coedited two volumes on the paleontology of Idaho, coauthored a volume on evolution and religion, and he was senior editor of *From Biped to Strider: The Emergence of Modern Human Walking, Running, and Resource Transport*. His involvement in the sasquatch question began after examining fresh tracks in southeastern Washington's Blue Mountains in 1996. He has assembled and analyzed the most significant collection of sasquatch footprint casts and photographs and has conducted collaborative fieldwork throughout the Pacific Northwest and Intermountain West under the moniker the North American Ape Project. His studies of this subject have been presented at numerous professional conferences and other public appearances. He is married to Terri Meldrum, and together they have six sons.